Extreme Weather Events and Vulnerability of Older People

Md Isahaque Ali · Dharmalingam Arunachalam ·
Helen Forbes-Mewett

Extreme Weather Events and Vulnerability of Older People

Experiences from Bangladesh and Other Asian Countries

Springer

Md Isahaque Ali
School of Social Sciences
Monash University
Clayton, VIC, Australia

Dharmalingam Arunachalam
School of Social Sciences
Monash University
Clayton, VIC, Australia

Helen Forbes-Mewett
School of Social Sciences
Monash University
Clayton, VIC, Australia

ISBN 978-3-032-11255-2 ISBN 978-3-032-11256-9 (eBook)
https://doi.org/10.1007/978-3-032-11256-9

This Springer imprint is published by the registered company Springer Nature Switzerland AG
The registered company address is: Gewerbestrasse 11, 6330 Cham, Switzerland

To our parents, teachers and family.

Preface

Bangladesh is one of the most disaster-prone areas in the world. Extreme weather events such as cyclones and floods have been extensively increasing in this area due to climate change, adversely affecting people and making them vulnerable to living and socioeconomic conditions. Among age groups, older people are the most vulnerable and have a lower adaptive capacity to cope with severe climate disasters and their impact on their socio-economic conditions. However, despite their vulnerable conditions to extreme weather events, there is relatively little knowledge on the vulnerability of older people to extreme weather events in Bangladesh. Some studies address the problems of older people in Bangladesh, but there has been no holistic study on their vulnerability to extreme weather events. This book fills this gap by providing a comprehensive exploration of socioeconomic and living vulnerabilities experienced by older people in a coastal region of Bangladesh. Besides exploring the vulnerability of older people to extreme weather events in the coastal area of the country, the book also examines the role of social protection programmes in addressing their vulnerabilities. The book investigates these issues through the experiences of 56 older people, both males and females, in a particular coastal area, Patharghata Upazila in Barguna district, in Bangladesh.

The book shows that older people in the coastal region of Bangladesh have been highly vulnerable to extreme weather events in the past, and they believe that the same or more brutal conditions will recur due to their poor socioeconomic condition and a lack of a proper support system. Every extreme event, especially cyclones and floods due to storm surges, destroyed or significantly damaged their livelihoods and negatively affected their lives, making them vulnerable. It further reveals that social protection programmes (formal and informal) help them to better cope with the impacts of extreme weather events and reduce their vulnerability.

This book presents the experiences of older people in a coastal region in Bangladesh in several chapters. Chapter 1 provides the background and context of the book with a discussion of the geographical and socioeconomic conditions of the country and outlines the objectives and the scope and limitations of the book. Chapter 2 investigates the vulnerability of older people in select Asian countries. Chapter 3 critically analyses the key concepts related to vulnerability and social protection and how together they determine the experiences of older people. Chapter 4 presents how the geographical location and socioeconomic factors escalate the vulnerabilities of older people and their families in the area. The livelihood and asset vulnerabilities of these people to extreme weather events are presented in Chapter 5 and the critical connection between extreme weather events and food security challenges among older people is presented in Chapter 6. The social protection measures that mitigate their vulnerabilities in different ways are presented in Chapter 7. The concluding chapter synthesises the experiences presented in the earlier chapters and discusses the policy implications.

This book is a valuable resource for graduate and postgraduate students, teachers, researchers, policymakers, and practitioners engaged in climate change, extreme weather events, vulnerability in old age, the role of social protection, Environmental Sociology, Social Work, Disaster Management, and Social Care. The book offers insights into the various socioeconomic challenges older people face in disaster-prone areas. It emphasises the necessity of social protection interventions to enhance their resilience and well-being. By illuminating the connections among aging, extreme weather events, and social protection, this book contributes to the broader discourse on sustainable development and the welfare of vulnerable populations during extreme weather occurrences. We hope this research will inspire additional investigations and actions to address the vulnerability of older people to climate change and extreme weather events, ultimately leading to more inclusive and resilient societies through social protection programmes.

Melbourne, Australia

Md Isahaque Ali
Dharmalingam Arunachalam
Helen Forbes-Mewett

Competing Interests The authors have no competing interests to declare that are relevant to the content of this manuscript.

Contents

Abbreviations

ASP	Adaptive Social Protection
ASRSP	Adaptive and Shock-Responsive Social Protection
BBS	Bangladesh Bureau of Statistics
BDT	Bangladeshi Taka
CEEW	Council on Energy, Environment and Water
DRFI	Disaster Risk Financing and Insurance
DSWD	Department of Social Welfare and Development
G2P	Government to Person
GFDRR	Global Facility for Disaster Reduction and Recovery
HIES	Household Income and Expenditure Survey
IPCC	Intergovernmental Panel on Climate Change
NGOs	Non-Governmental Organisations
OPM	Oxford Policy Management
PRECIS	Providing Regional Climates for Impact Studies
SDGs	Sustainable Development Goals
UK aid	United Kingdom aid
UNDP	United Nations Development Programme
UNDRR	United Nations Office for Disaster Risk Reduction
UNICEF	United Nations Children's Fund
USB	Universal Serial Bus

Chapter 1
Introduction

The world faces the intertwined challenges of ageing populations and extreme weather events, which are becoming more frequent and intense. The demographic shift, marked by an increasing number of older adults, is particularly pronounced in developing nations like Bangladesh, where climate change and existing socioeconomic vulnerabilities complicate the issues associated with ageing. This book, "Older People in Bangladesh: Extreme Weather Events, Social Protection, and Vulnerability, "addresses a crucial knowledge gap regarding these intersecting crises by examining the unique challenges faced by older people in this highly vulnerable environment in a comprehensive and timely manner. The book provides an in-depth analysis of how extreme weather events worsen the pre-existing vulnerabilities of older Bangladeshi citizens, as well as a critical examination of how social protection mechanisms can play a vital role in fostering resilience and mitigating these challenges.

Bangladesh is situated in the delta region of the Ganges-Brahmaputra-Meghna River system, which presents a higher risk of natural disasters. A tropical monsoon climate and low-lying topography render the area exceptionally susceptible to cyclones, floods, storm surges, riverbank erosion, and droughts. The impacts of climate change intensify these extreme weather events, which pose significant threats to the lives, livelihoods, and well-being of all Bangladeshis, particularly the elderly. Older individuals, who often face physical limitations, reduced mobility, and limited access to resources and information, are especially in vulnerable during and after disasters.

This book aims to provide a comprehensive and nuanced understanding of the numerous hazards faced by older people in Bangladesh, particularly in the context of recurrent extreme weather events. It goes beyond merely documenting these challenges and delves into the intricate interplay of factors contributing to these hazards. These factors encompass a broad range of issues, including but not limited to: geographic location and exposure to hazards; inadequate shelter and infrastructure that often struggle to endure the impact of extreme weather events; limited access to essential resources such as healthcare, clean water, sanitation, and transportation;

M. I. Ali et al., *Extreme Weather Events and Vulnerability of Older People*,
https://doi.org/10.1007/978-3-032-11256-9_1

pre-existing health conditions and disabilities that can be worsened by disasters, hindering an individual's capacity to cope and recover; the strength and scope of existing formal and informal social support systems; as well as the unique needs and experiences of older people, acknowledging that aging itself presents distinct challenges in disaster preparedness, response, and recovery. The book considers the diversity within the older population and recognises that vulnerability is further affected by factors such as gender, socio-economic status, place of residence, and disability.

A central and critical focus of "Older People in Bangladesh" is exploring social protection programmes and their potential to address the complex vulnerabilities of older people, considering climate change and environmental shocks. The book examines formal social safety nets, such as government-led initiatives and informal support systems within communities and families. It analyses the effectiveness of existing programmes while identifying gaps and weaknesses in their design and implementation. Additionally, it explores innovative approaches to strengthening social protection mechanisms, including cash transfers, social pensions, access to health care, and community-based disaster preparedness programmes, with a particular emphasis on their relevance to the needs of older people. The book underscores the urgent need for integrated and cross-sectoral strategies that consider the complex interplay of social, economic, environmental, and demographic factors affecting the lives of older people in Bangladesh.

Overall, "Older People in Bangladesh: Extreme Weather Events, Vulnerability and Social Protection" is a significant scientific resource and a compelling call to action. It offers vital data, thorough analysis, and evidence-based recommendations for policymakers, researchers, practitioners, development organisations, and anyone concerned about ageing, climate change, disaster risk reduction, and social protection issues. By emphasising the unique and often overlooked challenges faced by older people in Bangladesh, the book aims to raise awareness, encourage informed discussions, and promote the development of more inclusive, resilient, and equitable strategies for enhancing the well-being of older populations amid growing environmental and socio-economic challenges. This work is a timely and crucial addition to the expanding knowledge on the links between ageing, climate change, and vulnerability in the developing world, providing essential insights for fostering a more equitable and sustainable future for all.

1.1 Background to this Study

Bangladesh is a disaster-prone area in South Asia, with nearly 80% of the population exposed to floods and over 70% to cyclones (Acaps, 2019; Saha, 2017; World Bank, 2018). The country is vulnerable and susceptible to the impacts of extreme weather events due to its geophysical location, poor socioeconomic conditions and weak governance system (Alauddin & Rahman, 2013; Chowdhury et al., 2021). The country is a remarkably flat, low-lying alluvial plain crisscrossed by over 230 rivers and streams, with nearly 580 km of exposed coastline along the Bay of Bengal. People live in the low-lying coastal areas of the Ganges-Brahmaputra-Meghna delta, which covers most of the country and is subject to regular cyclones and floods (Uddin et al., 2019). Bangladesh is one of the top 10 countries most affected by climate change-induced natural disasters (Eckstein et al., 2019). Referring to the modelling data of Providing Regional Climates for Impact Studies (PRECIS), Bangladesh Economic Review 2019 stated that the annual average rainfall of Bangladesh would increase by about 4%, 2.3% and 6.7% by 2030, 2050 and 2070, respectively. The average temperature in Bangladesh has been rising and will rise even faster in the future (Rouf et al., 2011; Tanner et al., 2007). Since 1971, the average annual temperature increase has been 0.2 degrees Celsius per decade, well above the global rate of 0.13 degrees Celsius per decade (Rahman & Lateh, 2017). Another study based on 100 years of data found that the maximum temperature in Bangladesh has increased at a rate of 0.3 to 0.4 degrees Celsius per decade (Khatun et al., 2016). The average annual temperature in the country is expected to rise by about two degrees Celsius by 2050. A World Bank study found that up to two-thirds of Bangladesh is inundated by floods every three to five years, and 60% of the deaths from cyclones worldwide in the last 20 years have occurred in Bangladesh (World Bank and GFDRR, 2011). Besides, low-lying coastal areas are also at a higher risk of being subject to cyclones and storm surges. The Intergovernmental Panel on Climate Change (IPCC) predicts that by 2050, Bangladesh will lose 17% of its land and 30% of its food production because of the adverse impact of extreme events (Ministry of Finance, 2019).

Bangladesh ranks sixth among countries whose population is exposed to extreme hazards such as cyclones, droughts and floods and second in flood risk (Bernard et al., 2022; Christenson et al., 2014). Every year, on average, Bangladesh is hit by one extreme cyclone, 145 severe local convective storms and at least six tornadoes (Alam et al., 2003; Ahammad et al., 2013; Quader, 2017; Masum, 2019). Between 1960 and 2007, about 18 significant cyclones hit the Bangladesh coast, killing about 718,000 people (Haque et al., 2012). The Bhola cyclone of 1970, which killed up to 300,000 people, was the deadliest tropical cyclone in Bangladesh. The 1991 cyclone killed at least 138,000 people and up to ten million more were displaced. Almost 8.3 million people live in cyclone-prone areas, and this number is expected to increase to 20.3 million by 2050 (Tanner et al., 2007). Cyclones in Bangladesh cause death and destruction of livelihoods, as well as damage to crops, infrastructure, communities and communication networks (Haque & Jahan, 2016; Shaw et al., 2013). Cyclone SIDR in 2007 hit the southwest coast of Bangladesh with an average wind speed of 223 km/h, caused 3406 casualties, 55,282 injuries and affected 8.9 million people. It destroyed 0.6 million hectares of farmland, devastated 537,775 homes and flooded 8075 kms of roads. Damage and losses from cyclone SIDR were estimated at US$ 1.67 billion (GoB, 2008).

Storm surges are another extreme weather event in Bangladesh that occurs every year due to cyclonic events. Bangladesh is hit by two-fifths of all storm surges in the world. By 2050, cyclone-induced storm surges will inundate an additional 15% of Bangladesh's coastal area due to an expected 62 cm rise in sea level (World Bank, 2010). According to one study, the sea level in the coastal zone of Bangladesh is rising by 6–21 mms every year (Cell, 2016). By 2050 and 2100, a sea level rise of 25 cm and 1.0 m is projected to affect 4% and 17.5% of Bangladesh's total land mass, respectively (World Bank, 2000). Almost half of Bangladesh's coastal areas are currently at risk of cyclonic storm surges of more than one metre in height (Sarker & Ahmed, 2015). Furthermore, 60% of the country's total land area is affected by at least one extreme flood event every decade. The low-lying coastal regions of Bangladesh are prone to flooding and strong tidal effects that can spread 200 kms upstream from the coast (Dasgupta et al., 2010; Jisan et al., 2018). Flooding at different times of the year increases the vulnerability of people in the affected regions of the country. In the 1988 flood, 61% of the entire country was flooded; and in the 1998 flood, almost 68% of the country was flooded for days, affecting 45 million people, killing over 2300 people and damaging property worth $12 billion (Hofer & Messerli, 2006; Liong et al., 2000; Younus, 2014).

Riverbank erosion is considered the most dangerous hazard in Bangladesh (Alam & Collins, 2010; Penning-Rowsell et al., 2013). Riverbank erosion affects about 20 of the 64 districts in the country and approximately 200,000 people are displaced every year due to riverbank erosion (Aktar, 2012). Erosion of riverbanks has far-reaching consequences for the affected coastal regions and the country. People lose their land and belongings, assets and resources and become homeless due to riverbank erosion, resulting in displacement. As a result, people who become homeless have difficulty reintegrating into their new community, face social prejudice and suffer a severe lack of family and social ties (Hamide, 2017).

Droughts, heat waves and cold waves are caused by extremely high or low temperatures that put people at risk of health problems in different parts of the country. The frequency of droughts has increased by at least 10% in North Bengal, Bangladesh, between 1979 and 2018. Farmers will be forced to grow crops with lower economic yields, which may increase the vulnerability to food insecurity in the near future (Ahmed, 2006). Heat waves in Bangladesh are characterised by a rise in temperature and the absence of the usual pre-monsoon rains. A heat wave develops over Bangladesh when the maximum temperature rises above 36 °C. During summer, the maximum temperatures in Bangladesh range between 36 and 40 degrees Celsius (Masum, 2019). Bangladesh experienced 39 heat wave days (13 separate heat waves) in 23 years, from 1989 to 2011. During heat waves in Bangladesh, mortality increases disproportionately among older people and children (Bhuyan et al., 2018; Burkart et al., 2014). During the winter, cold waves in Bangladesh are accompanied by a rapid drop in the lowest temperature and constant cold winds from the Himalayas that affect the northern and northwestern parts of Bangladesh. In the winter of 2013, Bangladesh was hit by a cold wave that affected more than two million households. Homeless people, poor people without adequate winter clothing, the elderly, children and infants suffer the most during a cold wave (Masum, 2019).

The impacts of extreme weather events on the people, environment, ecosystems and infrastructures of affected communities in Bangladesh are devastating. The broader impacts of climate change and extreme weather events on resource systems and the environment can be considered, but ultimately, the main impacts are felt in the lives and livelihoods of marginalised groups, particularly older people living in the affected area (Toufique & Yunus, 2013). It has been shown that physical and cognitive factors, such as reduced physical health, mobility problems, cognitive decline and the influence of social factors, such as reduced social networks and socio-economic changes, make older people more vulnerable to extreme weather events (Fernandez et al., 2002). It is argued that there is a link between getting older and the likelihood of developing needs, which increases overall vulnerability. These changes in older people may affect their ability to respond effectively to a disaster, with the required response possibly beyond their coping capability (Pekovic et al., 2007). Disasters caused by severe weather undoubtedly have an enormous impact on older people. They affect their physical and mental health, infrastructure as well as the social cohesion of marginalised communities (Doherty & Clayton, 2011). Thus, older people are one of the most vulnerable groups, who are physically, financially, socially and even emotionally less capable of coping with repeated extreme weather events and socioeconomic crises (Haq et al., 2008).

1.2 Older People and Vulnerability in Bangladesh

The world is facing a continuous shift in the age structure of its population due to increasing life expectancy and decreasing fertility. People are living longer, and both the proportion and number of older people in the total population in Bangladesh are increasing significantly. The older population in Bangladesh was around 7.4% in 2010 (BBS, 2011). However, in 2019, about 13 million people in Bangladesh were over 60 years old, which is about 8% of the total population. With 36 million people over 60, the proportion of older people is expected to triple to 21.9% by 2050, which means that one in five Bangladeshis will be over 60 (Rahman et al., 2021; United Nations, 2019). Among the older population, a greater percentage are widowed (women 80% and men 39%).

Older people suffer from declining physical and mental abilities, dependency within the household and lack of access to regular income, employment and health care (Seplveda, 2010). In Bangladesh, 68% of the population aged over 60 are economically inactive, among which 95% are women, whereas 23% are self-employed in the agricultural (16%) and non-agricultural (7%) sectors (BBS, 2015). It is observed that the elderly who are not economically active may support themselves with the assets they accumulate over their life course (savings, rent from homeownership, land) pensions or in other ways (through assistance from family members, relatives and government and non-government agencies) of support and care (Rank et al., 2014; Rank & Hirschl, 2010; Plagnol, 2011). Most of the elderly live in single (widowed, divorced, separated, as well as never-married living separately) households (89%) and only 11% live in joint family households (living with sons or daughters). Single households generally consist of fewer members than joint family households. As a result, the elderly in single households often get less care than their counterparts who live in joint family households, where the possibility of caring for the elderly member is greater because of the presence of more family members (Bangladesh Bureau of Statistics, 2015).

The vulnerability of older people in Bangladesh is manifold. Older people in the country face difficulties in earning a consistent income as it becomes more difficult to find work as they age, especially in typical and physically demanding jobs. As the livelihood support and social security for older people in Bangladesh are inadequate, they are forced to look for work. Older people in the country also suffer from both degenerative and non-communicable diseases. For example, the osteoarthritis prevalence among the participants aged 60 to 69 significantly increased from 45% in 2020 to 55% in 2022 (Mistry et al., 2022). Again, the overall prevalence of multimorbidity among the elderly was 56.4% and the prevalence was higher among women (64%) than men (54%). The most common conditions were hypertension (33%), diabetes (27%), ischaemic heart disease (12%) and 9% were chronic obstructive pulmonary disease (Sara et al., 2018). Another study found that of the 65 participants, 88% had one or more non-communicable chronic diseases (Taskin et al., 2014). The main non-communicable diseases were hypertension (57%), arthritis (45%), diabetes mellitus (32%), stroke (20%) and myocardial infarctions (12%). In sum, older people in Bangladesh face various types of vulnerabilities that include physical vulnerabilities, lack of social dignity, economic crises, housing difficulties, illness, physical assault, mobility problems, family stresses, and food crises (Alam, 2015).

1.3 Social Protection in Bangladesh

Though the government's pension scheme was the only financial support programme for older people until the late 1990s, 114 programmes under 25 ministries are now implemented. These include allowances for the elderly, allowances for widows, allowances for destitute women, and payments for the poor and the disabled. It also includes the Employment Generating Programme for the Poorest, Food for Work, Work for Money, Test Relief and Vulnerable Group Feeding run by the Ministry of Disaster Management and Relief, and education and health incentives for the poor and vulnerable groups. The government pension programme covers only 5% of the workforce and 1% of older people in Bangladesh (World Bank, 2019; GED, 2019).

The Government of Bangladesh introduced the Old Age Allowance programme in 1997–98 to provide financial assistance for older people who had no access to the formal pension scheme. Through the Old Age Allowances programme, 44 million older people are now benefitting, each receiving Bangladeshi Taka (BDT) 500 (US$6) per month (Ministry of Social Welfare, 2019). In addition, about 17 million widows and deserted destitute women are receiving BDT 500 (US$6) per month as cash benefits through the Widow Allowance. The objectives of these programmes are to enhance the socio-economic conditions of the beneficiaries, improve their status in the family and society, and provide better nutrition and health in a sustainable way (Ministry of Social Welfare, 2019; World Bank, 2019). Social protection – materials and in-kind assistance – provided by the government and non-government organisations, and the support provided by family members, friends, community and society at large play a significant role in improving the socioeconomic conditions and mitigating vulnerabilities of older people to extreme weather events in coastal Bangladesh.

1.3.1 Old Age Allowance

The Old Age Allowance programme was introduced in fiscal year 1997–98 to provide social security to the country's elderly poor and low-income or income-eligible elderly and enhance the dignity of the family and society by the Department of Social Services of the Ministry of Social Welfare. Initially, 10 poor elderly people, including five men and five women, in each district of all Union Parishads in the country were provided with a grant of BDT 100 per month. Later, all parishes and municipalities in the country were included in this programme. To include more women in the programme, the age for women was lowered from 65 to 62, local MPs and other public representatives were included in the selection of beneficiaries, initiatives were taken to establish a database, and the money was disbursed by opening a bank account in the name of all beneficiaries upon payment of BDT 10. Since the 2017–2018 fiscal year, the old-age assistance programme has been partially handled through agent banking. Currently, all beneficiaries receive the monthly allowance in G2P (Government to Person) through mobile financial service providers, Nagad and bKash.

1.3.2 Allowance for the Widow and Husband Deserted Women

The allowance for widows and women abandoned (a woman whose husband is dead and Husband Deserted means, a woman who is divorced from her husband or does not live with her husband or has no contact for two years or more for any other reason) was introduced in the financial year 1998–99 by the Social Services Department of the Ministry of Social Welfare. At that time, 403,110 women were paid BDT 100 as monthly assistance. The programme was transferred from the Ministry of Social Welfare to the Ministry of Women and Children in the financial year 2003–04. In order to achieve greater adaptability in the implementation of this programme, the government reincorporated this programme into the Ministry of Social Welfare in fiscal year 2010–11. This programme has been successfully implemented by the Department of Social Services. Since fiscal year 2017–2018, part of the widow and widower allowance for abandoned women has been paid through agent banking. Recently, allowances for all beneficiaries have been disbursed through the Government to Person (G2P) system through the Nagad mobile financial service, Bkash. In 2022–23, 2,475,000 women were expected to receive BDT 500 as a monthly allowance, and the total allocation was BDT 14,954,000,000. In the last two years, measures have been taken to make the programme more transparent and accountable and to make it acceptable to all sections of the population, such as the revision of the implementation manual, the involvement of local MPs and other public representatives in the beneficiary selection process, and the creation of a database.

1.4 Significance of this Book

Bangladesh, a developing country in South Asia, is home to 168 million people. With 1116 people per square kilometre, it is one of the eight most populous countries in the world (World Population Review, 2022). Demographic changes in Bangladesh have resulted in an ageing society with many associated socio-economic and cultural challenges (Islam, 2016; Streatfield & Karar, 2008). The number of people aged 60 and older is increasing faster than the growth of the overall population. According to the Bangladesh Bureau of Statistics (BBS), the number of older people in Bangladesh is expected to reach about 20% (over 40 million) by 2050 (BBS, 2015). Though this growth of older people indicates human development and a significant increase in life expectancy, they are one of the vulnerable groups in the country (Hashemi, 2012).

The older population in Bangladesh is facing various social and economic problems including the break-down of the traditional family, change in social norms and values, unemployment and income insecurity, high level of medical cost, absence of a strong social security system and the indifference of implementing the elderly-friendly laws and regulations that increase their vulnerabilities (Barikdar et al., 2016). A study by the Bangladesh Bureau of Statistics (BBS) found that 32% of the country's older people lived below the poverty line, of which 11% lived in extreme poverty (BBS, 2016). Studies showed that about half of the older people in rural Bangladesh suffered from chronic energy deficiency and nearly two-thirds (62%) were at risk of malnutrition (Alam, 2015; Ferdous et al., 2009). It was also found that more than half (56%) of the older people were economically vulnerable and 6% suffered from food insecurity. In contrast, one in four older people was socially neglected (Alam, 2015). Although economic insecurity was one of the major vulnerabilities among older people in the country, most (69%) of the older people in Bangladesh were living without access to a pension or other social security/social protection benefits.

In addition to these socio-economic vulnerabilities, older people are more susceptible to life-threatening hazards caused by extreme weather events such as cyclones, storms, floods and extreme temperatures (Pekovic et al., 2007; Brunkard et al., 2008). Due to its geography and low adaptive capacity, older people of Bangladesh are vulnerable to devastating weather events such as floods, cyclones and storm surges, tornadoes and riverbank erosion (Chowdhury et al., 2021; Rahim et al., 2018; Uddin et al., 2019). These extreme climate events, directly and indirectly, affect the socio-economic conditions of society and the livelihoods of marginalised groups, especially the older people in the country. Compared to other age groups, older people suffer disproportionately from malnutrition during extreme weather events due to food insecurity, loss of household assets, a reduction in livelihood opportunities, social isolation and limited access to services (HelpAge International, 2009, 2015). In addition, socio-economic factors such as poverty, lack of neighbourhood support, poor housing conditions, limited access to resources and social networks increase their vulnerability to extreme weather events in the country (Haq et al., 2008, 2010; Lindley et al., 2011).

Despite the severe impact of extreme weather events and socio-economic conditions on the livelihoods of older people, only a few studies have examined the vulnerability of older people to extreme weather events and their relationship with socio-economic factors in the coastal region of Bangladesh (Ahsan, 2010). Several studies have shown the negative impact of natural disasters on society as a whole globally, but few with specifics on the vulnerability of older people in developing countries like Bangladesh.

There are some studies on the socio-economic problems of the older population in Bangladesh, such as the demographics of older people (Kabir, 1994), the health status of older people (Atiqur, 2012; Kabir et al., 2013; Munira, 2014; Probaha, 2014; Roy, 2002), elderly abuse, financial support to older people (Rahman et al., 2009), begging among older people (Rahman, 2009) and the living and care arrangement of older people (Munsur et al., 2010). Other studies in Bangladesh examined the health problems of older people in the context of natural disasters (Haq & Ahmed, 2017; Kabir et al., 2016; Kabir & Khan, 2014, 2016; Rahman, 2008a, 2008b; Ray-Bennett et al., 2010; Shahid, 2010). One recent study examined the vulnerability and adaptation strategies of aged people to cyclones in the coastal region of Bangladesh (Malak et al., 2020). It found that the vulnerability and capacity of older people depend on their socio-economic conditions, adaptive skills and available services.

Although social protection can contribute to reducing the vulnerability of older people to extreme weather events, there has not been any attempt to study the role of social protection in addressing older people's vulnerability to extreme weather events in Bangladesh. Therefore, researchers, policy experts and government officials in Bangladesh have been looking at how to address these vulnerabilities. Considering the precarious situation of the elderly population, the Government of Bangladesh has implemented some policies and programmes for their socio-economic development.

However, the vulnerability of older people is considered a complex phenomenon, and the consequences of extreme weather events and socioeconomic crises interact with biological processes, and behavioural, social, economic and political contexts throughout life (Bayraktar & Dal Yilmaz, 2018; Haq et al., 2008, 2010). Thus, this book explores the vulnerabilities of older people and the socio-governmental support system and its practical application concerning the needs of older people in the aftermath of extreme weather events in Bangladesh. This study explores the influence of extreme events on the vulnerability of older people in Bangladesh with a view to helping policymakers and researchers in this field to formulate a sustainable future action plan to alleviate the socio-economic vulnerabilities of the elderly in developing countries like Bangladesh.

To achieve the United Nations Sustainable Development Goals and ensure a secure, equitable and sustainable future, two major policy challenges - an ageing population and a changing climate - must be properly addressed (Haq, 2017). This book explores the intersection of three themes related to the vulnerability of older people - 'sources of vulnerability', 'factors exacerbating vulnerability to extreme weather events' and 'the role of social protection' in addressing the vulnerability of older people in coastal regions in coping with the impacts of extreme climate events. The findings have shown that there are strong links between resources, the impacts of extreme weather events and a lack of social protection that make older people in the region vulnerable. By considering the three interrelated issues together, this book contributes to the existing body of knowledge on the vulnerability of older Bangladeshis who are exposed to frequent extreme weather events.

This book not only provides a deeper understanding of older people's vulnerability to extreme weather events but also evidence-based knowledge on what factors are behind the vulnerability of older people in coastal villages of Bangladesh. This work operationalises the link between vulnerability to extreme weather events and social protection as the capacity to recover from the impacts of extreme weather events in the specific socio-economic and geophysical context of coastal settlements, adding value to the existing body of knowledge in these areas.

By placing social protection at the centre, this book not only addresses the issue of vulnerability more holistically but also provides a deeper and more nuanced understanding of the nature, causes and experiences of vulnerability among older people affected by cyclones and floods in coastal Bangladesh.

1.5 Study Participants

This book is based on the life experiences of 30 men and 30 women aged 60–80 years. The participants were selected from Patharghata Union of Patharghata Upazila (sub-district) within Barguna district. About 50% of the participants were beneficiaries of formal social protection programmes. The study site - the Patharghata Union - was selected purposively due to its susceptibility to extreme weather events. After consultation with the Chairman of the Patharghata Union Council, seven out of 12 villages in the Union - Rohita, Padma, Tengra, Charlathimara, Badurtala, Haritana and Gaharpur - were selected for this study because these villages are located near the sea and rivers and are frequently and severely affected by extreme weather events (Fig 1.1).

Due to its geographical location in the coastal regions, Patharghata is frequently exposed to various extreme weather events. It is one of the most vulnerable coastal regions in Bangladesh. Dealing with extreme weather events is a daily concern for the people of Patharghata, as the changing seasons bring a variety of unpredictable threats and stressors to their lives and socio-economic conditions. Evidence shows that

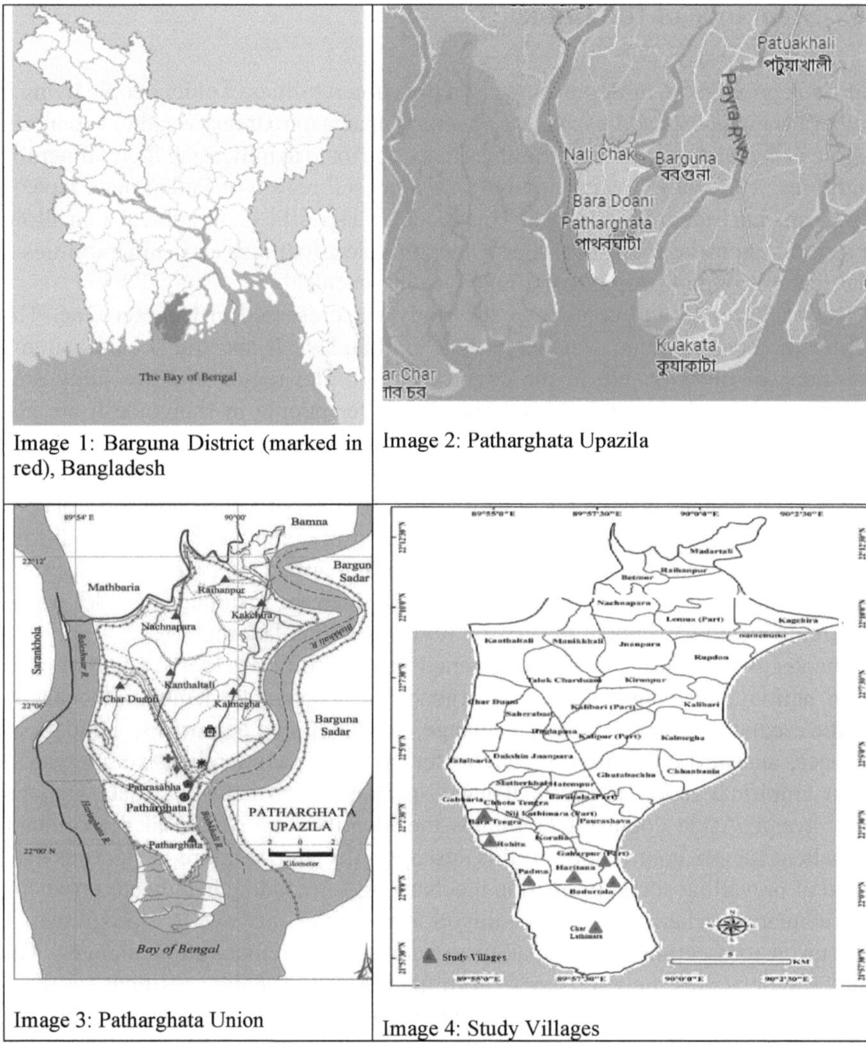

Image 1: Barguna District (marked in red), Bangladesh

Image 2: Patharghata Upazila

Image 3: Patharghata Union

Image 4: Study Villages

Fig. 1.1 Study location. *Source* https://commons.wikimedia.org/wiki/File:BD_Districts_LOC.svg

extreme weather events such as floods, cyclones, storm surges, salinisation, excessive fog, temperature fluctuations, erosion of riverbanks, heavy rains, a late rainy season and waterlogging are almost part of everyday life in the region, negatively affecting the lives and livelihoods of older people, their households and neighbours in Patharghata (Rahim et al., 2018).

Participants were interviewed for the study from December 2019 to March 2020. In addition, field notes were taken at the time of interviews and at the time of walking through different villages with the help of local guides to collect information beyond the interviews, in particular for data triangulation.

1.6 Structure of the Book

This book provides an in-depth analysis of the vulnerabilities of older people living in southern Bangladesh, addressing the adverse impacts of extreme weather events and the role of social protection programmes that contribute to mitigating their vulnerable conditions. The study was carried out in the coastal region of Patharghata Upazila of Barguna District in Bangladesh. This focus on a specific geographical area allows for a comprehensive analysis of the socioeconomic conditions and vulnerabilities of a particular group of people to extreme weather events.

The background and context of the study are presented in Chapter One. This chapter includes extreme weather events in Bangladesh and the socio-economic conditions of older people in the country, the coastal region and the study area. Social support policies and programmes for older people in Bangladesh are also presented in this chapter. This chapter analyses the nexus between older people and climate vulnerability in Bangladesh, focusing on Patharghata Union in Barguna district. This chapter argues for a reconceptualisation of vulnerability because of complex interactions between geographical exposure, socio-economic conditions, and inadequate protection systems. It argues for integrated interventions that extend beyond cash transfers to comprehensively strengthen the resilience of older people to extreme weather events in the coastal regions of Bangladesh.

Chapter two investigates the heightened vulnerability of elderly populations in Asian nations, attributed to the convergence of accelerated demographic ageing and the escalating effects of climate change, encompassing heatwaves, inundations, cyclones, and arid conditions. This chapter presents a comprehensive analysis of extant empirical research, national legal structures, and statistical data gathered from diverse Asian countries facilitates the identification of causal determinants and salient factors contributing to this susceptibility. Furthermore, a rigorous evaluation of prevailing social protection mechanisms—including healthcare provision, superannuation schemes, disaster mitigation measures, and welfare programs—is undertaken to ascertain their capacity to address these emergent challenges. This chapter underscores the exigency of formulating and executing resilient strategies for adaptive social welfare, thereby enhancing fortitude and safeguarding the aged demographic from extant and prospective risks stemming from demographic shifts and climatic perturbations.

Chapter three examines vulnerability as a multidimensional concept and analyses the physical, material, social, economic, and environmental factors that affect older people in different contexts. It establishes vulnerability as an inherently complex construct that requires integrated analytical approaches. This chapter outlines the evolution of vulnerability, demonstrating its transformation from a purely physical concept to a tuned understanding encompassing social and economic dimensions. This progression reflects a growing recognition of how various factors intersect to create distinctive vulnerability patterns among older people. It further provides a critical analysis of how older people's limited mobility and diminishing social support networks interact with broader societal structures, creating compound vulnerabilities that require targeted interventions. This chapter emphasises the importance of combining multiple perspectives of vulnerability to understand the vulnerabilities of older people thoroughly to extreme weather events. The chapter also evaluates social protection measures, including government interventions and community support strategies, assessing their effectiveness in reducing vulnerability and building resilience to extreme hazards. It covers the global, regional, and Bangladesh perspectives of extreme weather events and vulnerability of older people and the role of social protection. This chapter also presents an integrated framework to understand vulnerability to extreme events and social protection.

Chapter four demonstrates how geophysical location and settlement patterns fundamentally influence vulnerability among elderly coastal residents in Bangladesh. The analysis reveals that geographical positioning, particularly in low-lying coastal zones, creates inherent exposure to flooding and cyclonic activity. This physical vulnerability combines with socioeconomic factors to create complex risk patterns for older people and their families. Dwelling emerges as a critical concern, with many older people inhabiting houses constructed from substandard building materials. These structures offer minimal protection against extreme weather events, increasing physical risk and vulnerability during extreme climatic events. Besides, economic challenges manifest through various indicators, including available informal employment, limited land ownership, and persistent resource scarcity among the participants. Most older people in the area depend heavily on familial support, particularly from adult family members, creating potential instability in their support systems. This economic precarity directly influences their capacity to prepare for and recover from extreme weather events that frequently occur in the area. Social isolation of older people emerges as a significant factor, limiting access to care systems and reducing overall health resilience. Older people experience social dignity erosion, physical insecurity, and mobility challenges. These factors contribute to increased stress levels and reduce their coping capacity to withstand during and after extreme weather events. The chapter shows how geographic and socio-economic vulnerabilities create a distinctive convergence of risk factors among older people and their families in coastal Bangladesh.

Chapter five discusses the significant disruptions to economic resilience among older people and their families in extreme weather events in coastal Bangladesh. Cyclones and floods systematically reduce employment opportunities while simultaneously increasing household expenses, creating substantial financial pressure among them. This economic strain forces many people to rely on formal and informal debt mechanisms, potentially trapping them in cycles of financial vulnerability. Land and asset ownership emerges as a critical factor in determining resilience to extreme weather events. Limited property ownership reduces economic security among older people. This vulnerability becomes particularly acute during post-disaster recovery periods when assets prove crucial for rebuilding livelihoods. Housing experiences show persistent inadequacies against severe weather conditions. Many dwellings occupied by older residents lack fundamental structural resilience, leading to frequent displacement during extreme weather events. Older people face specific challenges during evacuations to cyclone shelters, highlighting accessibility issues and inadequate facilities for elderly residents. Older people often face difficult choices between maintaining their homes and investing in income-generating activities. The competing demands on the limited resources reduce overall resilience to extreme weather events.

Chapter six demonstrates critical connections between extreme weather events and food security challenges among older people in coastal Bangladesh. Cyclones and floods significantly disrupt established proper food management systems, affecting both accessibility and supply chain stability. These disruptions create particular challenges for older people, who often face reduced mobility and limited access to alternative food sources. It also identifies specific patterns where age-related health conditions compound the impacts of extreme weather events. Limited access to healthcare services during and after disasters creates additional risks for older people managing chronic psychophysical health conditions. There were systematic inadequacies in sanitation and hygiene infrastructure, particularly affecting older populations during extreme weather events. These deficiencies create increased health risks and compound their existing vulnerabilities. The limited governmental and non-governmental support networks struggle to address these vulnerabilities effectively. Food security emerges as a fundamental determinant of health outcomes among older people in the area. Older people in coastal regions experience chronic food insecurity which creates long-term health implications that reduce their capacity to cope with catastrophe events. Overall, this chapter argues how food security and health challenges create compound vulnerabilities among older people to extreme weather events.

Chapter seven chapter explores how social protection mechanisms reduce vulnerability among older people in coastal Bangladesh. Different formal social protection programmes, particularly the Old Age Allowance and Allowances for Widows and Deserted Women, have significant positive impacts on beneficiaries' livelihood standards and the social status of older people. These measures enable household savings, facilitate necessary investments, and improve access to essential services. The government programmes are complemented by informal support networks including volunteer organizations, non-governmental entities, and community-based support structures. These networks provide vital assistance in extreme weather events, often responding more quickly than formal institutions. Social protection programmes enhance resilience through multiple pathways. The combined social protection approaches deliver comprehensive support through food provision, clothing distribution, housing assistance, and direct monetary aid. Financial assistance enables better preparation for extreme weather events, while social networks facilitate their faster recovery. The chapter argues for increased investment in both formal social protection programmes and community-based support networks that are essential for developing effective vulnerability reduction strategies for older people.

Chapter eight synthesises the multifaceted lived experiences of older people across geographic, socioeconomic, and health-related dimensions in coastal regions of Bangladesh. This comprehensive analysis reveals the complex interactions between physical exposure, social support networks, and economic stability in determining the overall vulnerability of older people to extreme weather events. It concludes by emphasising the critical need to provide enhanced protection for older people from the impacts of extreme weather events. This also includes specific policy recommendations emphasising enhanced social protection systems, infrastructure improvements, and the development of other necessary interventions. These recommendations cover both immediate crisis response needs and long-term resilience-building measures for them.

In sum, this book provides valuable insights into the vulnerabilities of older people in coastal Bangladesh and the role of social protection programmes in addressing these challenges. It offers a comprehensive understanding of the multifaceted impacts of extreme weather events on the lives of older people and the need for holistic interventions to enhance their resilience and well-being.

References

ACAPS. (2019). *Rohingya crisis: Cyclones background created*. Retrieved October 13, 2019, from. https://www.acaps.org/sites/acaps/files/products/files/20180326_npm_acaps_analysis_hub_thematic_report_cyclones_background_rohingya_crisis.pdf

Ahammad, R., Nandy, P., & Husnain, P. (2013). Unlocking ecosystem-based adaptation opportunities in coastal Bangladesh. *Journal of coastal conservation, 17*(4), 833–840.

Ahmed, A. U. (2006). Bangladesh: Climate change impacts and vulnerability—A synthesis. Climate Change Cell, Department of Environment, Comprehensive Disaster Management Programme, Government of the People's Republic of Bangladesh. https://www.preventionweb.net/files/574_10370.pdf

Ahsan, M. N. (2010). *Climate change and socioeconomic vulnerability: Experiences and lessons from south-western coastal Bangladesh*. Wageningen University Library.

Aktar, N. (2012). *Prediction of river bank erosion along the Jamuna, the Ganges, the Padma and the lower Meghna rivers in 2012*. Center for Environmental and Geographic Information Services.

Alam, E., & Collins, A. E. (2010). Cyclone disaster vulnerability and response experiences in coastal Bangladesh. *Disasters, 34*(4), 931–954.

Alam, M. M., Hossain, M. A., & Shafee, S. (2003). Frequency of Bay of Bengal cyclonic storms and depressions crossing different coastal zones. *International Journal of Climatology: A Journal of the Royal Meteorological Society, 23*(9), 1119–1125.

Alam, S. (2015). *'Elderly people' in Bangladesh: Vulnerabilities, laws and policies (1st Draft Report)*. Retrieved December 23, 2019, from https://www.academia.edu/22978303/Elderly_people_in_Bangladesh_Vulnerabilities_laws_and_policies_1_st_Draft_Report

Alauddin, S. M., & Rahman, K. F. (2013). Vulnerability to climate change and adaptation practices in Bangladesh. *Journal of SUB, 4*(2), 25–42.

Atiqur, R. (2012). Human rights situation of older persons in Bangladesh. *Bangladesh Journal of Geriatrics, 47*, 15–23.

Bangladesh Bureau of Statistics (BBS). (2011). *Household Income and Expenditure Survey (HIES) 2010*. Statistics Division, Ministry of Planning, Government of Bangladesh.

Bangladesh Bureau of Statistics (BBS). (2015). *Elderly population in Bangladesh: Current features and future perspectives*. BBS Statistics and Informatics Division (SID), Ministry of Planning, Government of the People's Republic of Bangladesh.

Bangladesh Bureau of Statistics (BBS). (2016). *Household income and expenditure survey 2016*. Bangladesh Bureau of Statistics (BBS).

Bangladesh Bureau of Statistics. (2015). Elderly population in Bangladesh: Current features and future perspectives (Population Monograph Vol. 4). Statistics and Informatics Division, Ministry of Planning, Government of the People's Republic of Bangladesh. http://203.112.218.65:8008/WebTestApplication/userfiles/Image/PopMonographs/elderlyFinal.pdf (Accessed 20 April 2020).

Bangladesh Bureau of Statistics. (2015). Elderly population in Bangladesh: Current features and future perspectives (Population Monograph Vol. 4). Statistics and Informatics Division, Ministry of Planning, Government of the People's Republic of Bangladesh. http://203.112.218.65:8008/WebTestApplication/userfiles/Image/PopMonographs/elderlyFinal.pdf (Accessed 20 April 2020)

Barikdar, S. M., Islam, M. R., Kabir, R., & Zaman, A. M. (2016). Health profile of the elderly in Bangladesh: A review. *Journal of Gerontology and Geriatric Research, 5*(1), 1000266.

Bayraktar, N., & Dal Yilmaz, Ü. (2018). Vulnerability of elderly people in disasters: A systematic review. *Turkish Journal of Geriatrics/Türk Geriatri Dergisi, 21*(3), 467–482.

Bernard, A., Long, N., Becker, M., Khan, J., & Fanchette, S. (2022). Bangladesh's vulnerability to cyclonic coastal flooding. *Natural Hazards and Earth System Sciences, 22*(3), 729–751.

Bhuyan, M. D. I., Islam, M. M., & Bhuiyan, M. E. K. (2018). A trend analysis of temperature and rainfall to predict climate change for northwestern region of Bangladesh. *American Journal of Climate Change, 7*(2), 115–134.

Brunkard, J., Namulanda, G., & Ratard, R. (2008). Hurricane Katrina deaths, Louisiana, 2005. *Disaster Medicine and Public Health Preparedness, 2*(4), 215–223. https://doi.org/10.1097/DMP.0b013e31818aaf55

Burkart, K., Khan, M. M. H., Schneider, A., Breitner, S., Langner, M., Kraemer, A., & Endlicher, W. (2014). The effects of season and meteorology on human mortality in tropical climates: A systematic review. *Transactions of the Royal Society of Tropical Medicine and Hygiene, 108*(7), 393–401.

Cell, C. C. (2016). Assessment of sea level rise on Bangladesh coast through trend analysis. Climate Change Cell (CCC), Department of Environment, Ministry of Environment and Forests.

Chowdhury, M. A., Hasan, M. K., & Islam, S. L. U. (2021). Climate change adaptation in Bangladesh: Current practices, challenges and way forward. *The Journal of Climate Change and Health, 1*, 100108.

Christenson, E., Elliott, M., Banerjee, O., Hamrick, L., & Bartram, J. (2014). Climate-related hazards: A method for global assessment of urban and rural population exposure to cyclones, droughts, and floods. *International Journal of Environmental Research and Public Health, 11*(2), 2169–2192.

Dasgupta, S., Huq, M., Khan, Z. H., Ahmed, M. M. Z., Mukherjee, N., Khan, M., & Pandey, K. D. (2010). *Vulnerability of Bangladesh to cyclones in a changing climate: Potential damages and adaptation cost* (World Bank Policy Research Working Paper 5280).

Doherty, T. J., & Clayton, S. (2011). The psychological impacts of global climate change. *American Psychologist, 66*(4), 265.

Eckstein, D., Hutfils, M. L., & Winges, M. (2019). *Global climate risk index 2019: Who suffers most from extreme weather events? Weather-related loss events in 2017 and 1998 to 2017*. German Watch.

Ferdous, T., Kabir, Z., Wahlin, Å., Streatfield, K., & Cederholm, T. (2009). The multidimensional background of malnutrition among rural older individuals in Bangladesh—A challenge for the millennium development goal. *Public Health Nutrition, 12*(12), 2270–2278.

Fernandez, L. S., Byard, D., Lin, C. C., Benson, S., & Barbera, J. A. (2002). Frail elderly as disaster victims: Emergency management strategies. *Prehospital and Disaster Medicine, 17*(2), 67–74.

General Economics Division (GED), Bangladesh Planning Commission. (2019). Social Security Digital Cash Transfers: Strengthening G2P in Bangladesh. Cabinet Division & GED. https://socialprotection.gov.bd/wpcontent/uploads/2020/07/Social-Security-Digital-Cash-Transfers-Strengthening-G2P-in-Bangladesh.pdf (Accessed 20 April 2020).

GoB. (2008). *Cyclone Sidr in Bangladesh: Damage, loss and needs assessment for disaster recovery and reconstruction*. Refrieved from https://reliefweb.int/sites/reliefweb.int/files/resources/F2FDFF067EF49C8DC12574DC00455142-Full_Report.pdf [Access date 26 October 2020].

Hamide, S. (2017). *A review study of literature relating to coastal disaster in Bangladesh*. Retrieved November 10, 2020, from https://www.researchgate.net/publication/327238156_A_review_study_of_literature_relating_to_coastal_disaster_in_Bangladesh

Haq, G. (2017). Growing old in a changing climate. *Public Policy & Aging Report, 27*(1), 8–12.

Haq, M. A., & Ahmed, S. (2017). Climate change and health vulnerabilities of older adults in Bangladesh. *Journal of Environmental Health Science & Engineering, 15*, 23.

Haq, G., Brown, D., & Hards, S. (2010). *Older people and climate change: The case for better engagement*. Stockholm Environment Institute.

Haq, G., Whitelegg, J., & Kohler, M. (2008). *Growing old in a changing climate: Meeting the challenges of an ageing population and climate change*. Stockholm Environment Institute.

Haque, U., Hashizume, M., Kolivras, K. N., Overgaard, H. J., Das, B., & Yamamoto, T. (2012). Reduced death rates from cyclones in Bangladesh: What more needs to be done? *Bulletin of the World Health Organisation, 90*, 150–156.

Haque, A., & Jahan, S. (2016). Regional impact of cyclone Sidr in Bangladesh: A multi-sector analysis. *International Journal of Disaster Risk Science, 7*(3), 312–327.

Hashemi, S. M. (2012). Social protection and older people. *Probeen Kantho, 8*. Resource Integration Centre. Retrieved June 12, 2019, from http://ricbd.org/archives/publications/

HelpAge International. (2009). *Witness to climate change: Learning from older people's experience*. HelpAge International. Retrieved July 22, 2020, from https://www.helpage.org/silo/files/witness-to-climate-change-learning-from-older-peoples-experience.pdf

HelpAge International. (2015). *Climate change in an ageing world*. HelpAge International. Retrieved July 22, 2020, from https://www.helpage.org/silo/files/climate-change-in-an-ageing-world.pdf

Hofer, T., & Messerli, B. (2006). Floods in Bangladesh: History, dynamics and rethinking the role of the Himalayas. *Ecology, 29*, 254–283.

Islam, M. M. (2016). Demographic transition and the emerging windows of opportunities and challenges in Bangladesh. *Journal of Population Research, 33*(3), 283–305.

Jisan, M. A., Bao, S., & Pietrafesa, L. J. (2018). Ensemble projection of the sea level rise impact on storm surge and inundation at the coast of Bangladesh. *Natural Hazards and Earth System Sciences, 18*(1), 351–364.

Kabir, M. H. (1994). *Local level policy development to deal with the consequences of population ageing in Bangladesh (No. 131)*. Economic & Social Commission for Asia and the Pacific (ESCAP).

Kabir, R., Kabir, M., Gias Uddin, M., Ferdous, N., & Khan Chowdhury, M. (2016). Elderly population growth in Bangladesh: Preparedness in public and private sectors. *IOSR Journal of Humanities and Social Science, 21*(8), 58–73.

Kabir, R., & Khan, H. T. (2014). Climate change and public health situations in the coastal areas of Bangladesh. *International Journal of Social Science Studies, 2*, 109.

Kabir, R., & Khan, H. T. (2016). Climate change impact: The experience of the coastal areas of Bangladesh affected by cyclones Sidr and Aila. *Journal of Environmental and Public Health, 2016*, 9654753.

Kabir, R., Khan, H. T., Kabir, M., & Rahman, M. T. (2013). Population ageing in Bangladesh and its implications on health care. *European Scientific Journal, 9*(33), 34–47.

Khatun, M. A., Rashid, M. B., & Hygen, H. O. (2016). *Climate of Bangladesh, MET Report*. Bangladesh Meteorological Department and Norwegian Meteorological Institute.

Lindley, S., O'Neill, J., Kandeh, J., Lawson, N., Christian, R., & O'Neill, M. (2011). Climate change, Justice and Vulnerability. Joseph Rowntree Foundation. https://www.climatejust.org.uk/sites/default/files/5.%20climate-change-social-vulnerability-full.pdf

Liong, S. Y., Lim, W. H., Kojiri, T., & Hori, T. (2000). Advance flood forecasting for flood stricken Bangladesh with a fuzzy reasoning method. *Hydrological Processes, 14*(3), 431–448.

Malak, M. A., Sajib, A. M., Quader, M. A., & Anjum, H. (2020). "We are feeling older than our age": Vulnerability and adaptive strategies of aging people to cyclones in coastal Bangladesh. *International Journal of Disaster Risk Reduction, 48*, 101595.

Masum, J. H. (2019). *Climatic hazards in Bangladesh: A literature review*. Coastal Development Partnership (CDP).

Ministry of Finance, Bangladesh. (2019). *Bangladesh Economic Review*. Retrieved August 21, 2019, from https://mof.gov.bd/site/page/44e399b3-d378-41aa-86ff-8c4277eb0990/Bangladesh-Economic-Review-Archive

Ministry of Social Welfare. (2019). *Annual Report 2018–19*. Government of the People's Republic of Bangladesh.

Mistry, S. K., Ali, A. R. M. M., Yadav, U. N., Das Gupta, R., Anwar, A., Basu, S., Huda, M. N., & Mitra, D. K. (2022). A tale of osteoarthritis among older adults during the COVID-19 pandemic in Bangladesh: A repeated cross-sectional study. *PLoS One, 17*(9), e0274838.

Munira, J. (2014). Elderly people and human rights: An overview of Bangladesh. *Bangladesh Journal of Geriatrics, 49*, 75–92.

Munsur, A. M., Tareque, I., & Rahman, K. M. (2010). Determinants of living arrangements, health status and abuse among elderly women: A study of rural Naogaon district, Bangladesh. *Journal of International Women's Studies, 11*(4), 162–176.

Pekovic, V., Seff, L., & Rothman, M. (2007). Planning for and responding to special needs of elders in natural disasters. *Generations, 31*(4), 37–41.

Penning-Rowsell, E., Sultana, P., & Thompson, P. (2013). The last resort: Population movement in Bangladesh under climate change. *International Journal of Disaster Risk Reduction, 3*, 1–8.

Plagnol, A. C. (2011). Financial satisfaction over the life course: The influence of assets and liabilities. *Journal of Economic Psychology, 32*(1), 45–64.

Probaha, B. (2014). Mental health of older persons: A neglected issue. *Bangladesh Journal of Geriatrics, 49*.

Quader, M. A., Khan, A. U., & Kervyn, M. (2017). Assessing risks from cyclones for human lives and livelihoods in the coastal region of Bangladesh. *International Journal of Environmental Research and Public Health, 14*(8), 831.

Rahim, M. A., Siddiqua, A., Nur, M. N. B., & Zaman, A. M. (2018). Community perception on adverse effects of natural hazards on livelihood and enhancing livelihood resiliency: A case study at Patharghata Upazila, Barguna. *Procedia Engineering, 212*, 149–156.

Rahman, A. (2008a). Climate change and its impact on health in Bangladesh. *The Regional Health Forum, 12*(1), 16–26.

Rahman, M. (2008b). Heat-related morbidity among children and older people during floods in Bangladesh. *International Journal of Environmental Research and Public Health, 5*(6), 270–277.

Rahman, F. M., Banu, S. J., & Akhter, N. (2009). Financial support patterns to older people in Bangladesh. *Journal of Social Welfare and Family Affairs, 31*(2), 105–118.

Rahman, F. N., Khan, H. T., Hossain, M. J., & Iwuagwu, A. O. (2021). Health and wellbeing of indigenous older adults living in the tea gardens of Bangladesh. *PLoS One, 16*(3), e0247957.

Rahman, M. R., & Lateh, H. (2017). Climate change in Bangladesh: A spatio-temporal analysis and simulation of recent temperature and rainfall data using GIS and time series analysis model. *Theoretical and Applied Climatology, 128*, 27–41.

Rank, M. R., Hadley, H. S., & Williams, J. H. (2014). A life course approach to understanding poverty among older American adults. *Families in Society: The Journal of Contemporary Human Services, 91*(4), 337–341.

Rank, M. R., & Hirschl, T. A. (2010). *Estimating the life course dynamics of asset poverty (CSD Working Paper No. 10–25)*. St. Louis, MO: Washington University, Center for Social Development. https://doi.org/10.7936/K7N0162W

Ray-Bennett, N. S., Collins, A., Bhuiya, A., Edgeworth, R., Nahar, P., & Alamgir, F. (2010). Exploring the meaning of health security for disaster resilience through people's perspectives in Bangladesh. *Health & Place, 16*(3), 581–589.

Rouf, M. A., Uddin, M. K., Debsarma, S. K., & Rahman, M. M. (2011). Climate of Bangladesh: An analysis of northwestern and southwestern part using high resolution atmosphere-ocean general circulation model (AOGCM). *The Agriculturists, 9*(1–2), 143–154.

Roy, S. (2002). Psychological problems of older people and some aspects of alighting from these problems. *Social Science Journal, Rajshahi University, 7*, 73–83.

Saha, C. K. (2017). Cyclone, salinity intrusion and adaptation and coping measures in coastal Bangladesh. *Space and Culture, India, 5*(1), 12.

Sara, H. H., Chowdhury, M. A. B., & Haque, M. A. (2018). Multimorbidity among elderly in Bangladesh. *Aging Medicine, 1*(3), 267–275.

Sarker, M. H., & Ahmed, F. (2015). *Climate change vulnerability of drinking water supply*. IUCN, International Union for Conservation of Nature.

Seplveda, M. (2010). *Report of the independent expert on the question of human rights and extreme poverty*. SSRN 2441170. United Nations Human Rights Council.

Shahid, S. (2010). Probable impacts of climate change on public health in Bangladesh. *Asia Pacific Journal of Public Health, 22*(3), 310–319.

Shaw, R., Mallick, F., & Islam, A. (Eds.). (2013). *Disaster risk reduction approaches in Bangladesh* (Vol. 103). Springer.

Streatfield, P. K., & Karar, Z. A. (2008). Population challenges for Bangladesh in the coming decades. *Journal of Health, Population and Nutrition, 26*(3), 261–272.

Tanner, T., Hassan, A., Islam, K. M., Conway, D., Mechler, R., Ahmed, A. U., & Alam, M. (2007). *ORCHID: Piloting climate risk screening in DFID Bangladesh*. Summary research report. Institute of Development Studies, University of Sussex.

Taskin, T., Biswas, T., Siddiquee, A. T., Islam, A., & Alam, D. (2014). Chronic non-communicable diseases among the elderly in Bangladesh old age homes. *The International Journal of Aging and Society, 3*(4), 67–75.

Toufique, K. A., & Yunus, M. (2013). Vulnerability of livelihoods in the coastal districts of Bangladesh. *The Bangladesh Development Studies, 36*(1), 95–120.

Uddin, M. N., Islam, A. S., Bala, S. K., Islam, G. T., Adhikary, S., Saha, D., & Akter, R. (2019). Mapping of climate vulnerability of the coastal region of Bangladesh using principal component analysis. *Applied Geography, 102*, 47–57.

United Nations. (2019). *World population ageing 2019: Highlights (ST/ESA/SER. A/430)*. Department of Economic and Social Affairs, Population Division.

World Bank. (2000). *Bangladesh: Climate change and sustainable development. Report No 21104 BD*. World Bank.

World Bank. (2010). *Economics of adaptation to climate change: Bangladesh*. World Bank.

World Bank. (2018). *Bangladesh disaster risk and climate resilience programme*. Retrieved September 14, 2019, from https://www.worldbank.org/en/country/bangladesh/brief/bangladesh-disaster-risk-climate-change-programme

World Bank. (2019). *Social protection and labor*. Retrieved September 30, 2020, from https://www.worldbank.org/en/topic/socialprotection

World Bank and GFDRR. (2011). *Vulnerability, Risk reduction, and adaptation to climate change: Bangladesh*. World Bank.

World Population Review. (2022). Bangladesh population 2022. Retrieved November 26, 2022, from https://worldpopulationreview.com/countries/bangladesh-population

Younus, M. A. F. (2014). *Vulnerability and adaptation to climate change in Bangladesh: Processes, assessment and effects*. Springer.

References

[illegible]

Chapter 2
Older People and Extreme Weather Events in Asia

2.1 Introduction

The Asian continent is experiencing a period of significant transformation characterised by the confluence of rapid population aging and the increasing severity of extreme weather events. This intersection has resulted in a complex risk environment wherein older adults demonstrate elevated and interconnected vulnerabilities to a range of climate-related hazards, including heatwaves, extensive flooding, intense tropical cyclones, torrential rainfall, severe storm surges, and prolonged droughts (Guanio, 2021; HelpAge International, 2024; Roy & Hulme, 2003). Projections indicate that Asia will be home to approximately 1.3 billion individuals aged 60 and over by 2050. This represents a threefold increase from 2010 levels, signifying that one in four inhabitants will be classified as elderly (Asian Development Bank, 2025; Kinsella & Phillips, 2005). This demographic shift is occurring at a particularly rapid pace in several nations, including China, Sri Lanka, Thailand, and Vietnam (Asian Development Bank, 2025; ESCAP, 2017; ESCAP, 2025). Concurrently, the region experiences a disproportionate burden of climate-induced hydrometeorological disasters (O'Brien et al., 2006; Stockholm Environment Institute, 2024; UNDRR, 2024). For example, the year 2023 saw 79 documented hydrometeorological hazards across the Asia-Pacific, directly affecting over nine million individuals and causing over 2000 fatalities (UNDRR, 2024). Preliminary analyses, dating back to the late 1990s, have underscored the specific vulnerabilities of aging populations in disaster contexts (Partridge et al., 2018).

This chapter presents a consolidated analysis of the vulnerabilities of older persons to extreme weather events across various Asian contexts. Employing an extensive review of scholarly research, policy documents, and irrelevant data from selected countries—namely, China, Japan, India, Bangladesh, the Philippines, Indonesia, Thailand, Vietnam, Nepal, Myanmar, and Pakistan—this chapter pursues three principal objectives. First, it aims to identify the key physiological, socio-economic, and

M. I. Ali et al., *Extreme Weather Events and Vulnerability of Older People*,
https://doi.org/10.1007/978-3-032-11256-9_2

institutional determinants that contribute to the varied vulnerability of older populations when facing hydro-meteorological hazards (Wisner et al., 2004). Second, it evaluates the present condition and inherent constraints of current social protection frameworks in mitigating extreme weather risks for older individuals within specific national environments (Barrientos & Hulme, 2008). Third, this chapter posits the critical importance of adaptive social protection in enhancing resilience and enabling efficient, age-inclusive responses to the effects of climate change (Heltberg et al., 2009). Accordingly, this examination establishes social protection not solely as a reactive safeguard but as an essential and proactive element of climate change adaptation and disaster risk reduction strategies, particularly within the framework of extreme weather events.

2.2 Contextual Manifestations and Adaptive Responses to Extreme Weather

The generalised vulnerabilities discussed previously manifest with distinct characteristics across individual Asian nations when subjected to extreme weather events. These variations are substantially determined by specific demographic profiles (Lutz et al., 2001), established socio-economic frameworks (Adger, 2006), and the effectiveness of governance mechanisms in managing hydro-meteorological hazards.

In **China**, a demographic trend of rapid aging correlates with increased mortality risks. These risks are directly attributable to rising ambient temperatures and acute heat wave phenomena (Sun et al., 2016; Yao et al., 2025). Vulnerability to the deleterious effects of extreme heat is complexly determined by age, pre-existing medical conditions, geographic context (urban versus rural exposures), and socioeconomic status (Meehl & Tebaldi, 2004; Yao et al., 2025). While national policy frameworks exist for climate-sensitive health adaptation and the implementation of early heat alert systems (China CDC, 2025), anecdotal evidence suggests a perceived deficiency in governmental aid during extended heatwave occurrences (Yang & Zhang, 2025). Additionally, China experiences recurrent and substantial flooding events, particularly in the vicinity of major fluvial systems such as the Yangtze River. Research indicates heightened risk profiles for older, geographically isolated rural populations during these flood episodes (Tian & Lemos, 2018). Quantitative data specifying the direct mortality or displacement effects of floods on the elderly demographic are occasionally less articulated in national summaries compared to data concerning heat impacts. Nonetheless, evaluations of the medical insurance paradigm reveal inequities in healthcare provision for elderly individuals facing broader climate-related hazards (Yang & Zhang, 2025), and identified systemic deficiencies in policies related to elderly services may undermine efficacious responses to diverse climate impacts, encompassing flood events (Yand and Zhang, 2025).

Japan, characterised by a super-aged society, faces considerable challenges from extreme weather events. Mortality among older persons due to heatwaves remains a significant concern (Japan Centre for International Exchange, 2024; Nakai et al., 1999; The Japan Times, 2018). In addition to thermal stress, major flood events, such as the 2018 Japan Floods, present substantial risks, notably impacting the Long-Term Care Insurance system by increasing costs and augmenting care requirements for elderly individuals displaced or otherwise affected by widespread inundation (Yang & Hayashi, 2023; Yoshida et al., 2022). The nation is also highly susceptible to typhoons, which frequently precipitate widespread flooding and landslides. Research following events such as Typhoon Tokage (2004) identified difficulties in evacuating older adults with mobility impairments. While the legal framework ostensibly prioritises older persons in disaster management and evacuation planning, the effective implementation of these measures during flood or typhoon events encounters persistent challenges. These challenges encompass difficulties in inter-agency coordination and a sometimes inadequate understanding of the specific needs of older adults within emergency shelters and during evacuation procedures (Japan Centre for International Exchange, 2024; Tamura et al., 2009). Non-governmental organisations (NGOs) often provide critical support in the aftermath of such weather-related disasters (GlobalGiving, 2011; Shaw & Takeuchi, 2005).

India is characterised by a high degree of susceptibility to extreme heat, with geographically delineated regions of elevated risk and pronounced effects on elderly populations, particularly those residing within urban heat islands (CEEW, 2025; Chatterjee et al., 2025; Mazdiyasni et al., 2017). In addition to thermal stress, India routinely experiences severe inundations, especially during the monsoon season, resulting in extensive infrastructural damage and population displacement. Elderly individuals frequently encounter amplified challenges in accessing relief services and medical care in the aftermath of such floods (Carter et al., 2025; Lam & Roy, 2019; Hashim & Hashim, 2016). The nation is also vulnerable to the impact of destructive tropical cyclones affecting its substantial coastal territories, wherein elderly members of fishing and agricultural communities are particularly exposed to adverse effects (Mohanty et al., 2020; Raghavan & Rajesh, 2003). While the National Action Plan for Older Adults addresses welfare concerns (Ministry of Social Justice and Empowerment, 2020), detailed, granular data within the aforementioned sources regarding the direct consequences of flood and cyclone occurrences, specifically concerning mortality or displacement among the elderly across diverse geographical locations, is less comprehensively presented when juxtaposed with data pertaining to heatwaves. Prevailing economic vulnerability, however, markedly diminishes the overall adaptive capacity of elderly individuals when confronted with climatic shocks (CEEW, 2025; Jha, 2014). Notably, heatwaves are currently designated as events eligible for dedicated financial support, which fosters enhanced preparedness mechanisms; nevertheless, a comparable extent of explicit, elderly-centric focus within disaster response protocols for floods and cyclones is not universally discernible across all national and sub-national strategies examined within the evaluated text (CEEW, 2025).

In the context of **Bangladesh**, a country highly susceptible to cyclones, floods, and river erosion, the accelerated aging of the populace alongside a relatively low-income level substantially increases the vulnerability of older adults (HelpAge International,

2020; HelpAge International, 2024; Huq et al., 1999). River erosion, often exacerbated by repeated flooding, precipitates considerable losses in agricultural land and livelihoods, disproportionately affecting elderly individuals who frequently rely on agricultural activities (HelpAge International, 2020; HelpAge International, 2024; Islam & Rashid, 2011). Bangladesh's considerable cyclone vulnerability, exemplified by occurrences such as Cyclone Sidr (2007), leads to extensive devastation and displacement, with older persons experiencing particular hardships due to physical constraints during evacuations and elevated health hazards in often overcrowded shelters (HelpAge International, 2008; Nadiruzzaman (2012)). The current social protection framework, including the Old Age Allowance, is significantly undermined by its meagre financial provision and a critical absence of climate-adaptive modifications required during regular flood and cyclone incidents. This deficiency restricts the capacity of older persons to recoup lost resources and reestablish their lives (Afrin et al., 2025; Hasan, 2012; HelpAge International, 2024). Notwithstanding the presence of broad disaster preparedness policies, the unique vulnerabilities faced by older persons during these hydro-meteorological events are frequently disregarded in planning and response measures (HelpAge International, 2024). Ongoing advocacy is focused on fostering universal pension programs better designed to mitigate weather-related disturbances and to guarantee the engagement of older persons in disaster management planning (HelpAge International, 2022; HelpAge International, 2024).

The Republic of the **Philippines** is highly susceptible to typhoons, experiencing an average of 20 tropical cyclones per annum (Bankoff, 2003; Guanio, 2021; UNDRR, 2024). These meteorological events, which frequently include floods and storm surges, pose substantial and continuous threats to the nation's expanding older demographic, a population rendered more vulnerable by physical frailty and limited financial resources (Guanio, 2021; Heijmans, 2004). The devastating impact of Typhoon Haiyan (Yolanda) in 2013 exemplified the disproportionate consequences for older individuals, including difficulties in obtaining assistance and meeting long-term recovery requirements (Reyes et al., 2018; Almazan et al., 2019). Extreme meteorological disruptions have the potential to precipitate household poverty, thereby affecting older members within these households (Guanio, 2021). The national development plan and the Adaptive and Shock-Responsive Social Protection (ASRSP) roadmap are designed to enhance the responsiveness of social protection strategies to these recurring weather catastrophes. A key measure is the Emergency Cash Transfer program, which provides essential financial aid to those affected by typhoons (DSWD, 2021; National Economic and Development Authority, 2017). Legal provisions mandating resilient evacuation centres explicitly include considerations for older persons, recognising their unique requirements during storm-related evacuations (Refworld, 2024; The Judiciary PH, 2024). However, ensuring the effective implementation of these policies for the most disadvantaged elderly individuals amidst widespread typhoon impacts remains a significant challenge (Guanio, 2021).

Indonesia faces substantial climate-related vulnerabilities, encompassing recurring floods and droughts, which negatively influence agricultural productivity and increase food insecurity. Elderly Indonesian individuals, especially those residing in geographically exposed regions such as low-lying coastal areas or riverine settlements, are subject to climate effects that disrupt conventional farming and fishing occupations, inherently susceptible to variations in precipitation patterns and sea-level elevation. Although a significant risk of heat-related stress for older persons is acknowledged, inundation episodes, prevalent throughout the archipelago, frequently cause community displacement and impede access to indispensable utilities, with older persons often undergoing extended recuperation durations. The nation is strategically endeavouring to incorporate Adaptive Social Protection (ASP) into its national development strategies to holistically confront these climate risks, inclusive of those impacting the elderly. This strategic incorporation entails the exploration of innovative fiscal mechanisms, such as Disaster Risk Financing and Insurance (DRFI), for the expeditious provision of emergency social protection following climate-related incidents. Nonetheless, the efficacious operationalisation of ASP at the local administrative level to specifically mitigate weather impacts such as floods, particularly for marginalised elderly demographics, encounters practical implementation obstacles (Juliannisa et al., 2025).

Thailand is highly vulnerable to hydro-meteorological hazards, with floods, landslides resulting from heavy rainfall, and severe storms being primary concerns. The 2011 floods demonstrated significant vulnerabilities among elderly urban and rural populations. Older persons in Thailand face complex vulnerabilities as age-related frailty coincides with adverse environmental conditions, including heat waves and cold snaps (Poapongsakorn & Meethom, 2013). They are often financially disadvantaged, limiting their ability to take protective measures against extreme weather events. While existing social protection frameworks theoretically allow for vertical expansion during emergencies, the practical delivery of aid during floods can be impeded by systemic fragmentation and logistical obstacles. Outdated beneficiary databases and digital literacy limitations among older persons seeking assistance exacerbate these issues. Criticisms have identified ongoing challenges of poor inter-agency coordination and inadequate early warning systems for floods. Although temporary adjustments to social security provisions have been made following recent disasters, there is a discernible need for more permanent, automated response mechanisms linked to predefined disaster triggers to ensure timely and adequate support for older persons (Khan & Marzi, 2019, May).

Vietnam is significantly affected by climate change and experiences increasing natural disasters, exhibiting high vulnerability to typhoons, floods, and droughts, particularly in the Mekong Delta and coastal areas. This escalating risk coincides with rapid population aging, imposing a disproportionate burden on older individuals. Extreme weather events pose a substantial threat to older persons due to existing health issues and inadequate age-friendly infrastructure for safe evacuation and shelter. The considerable macroeconomic costs and elevated poverty risks resulting from these events are exacerbated by adverse effects on agriculture and critical infrastructure. While Vietnam has made progress in developing its social protection framework, including integrated approaches for vulnerable demographics, challenges persist concerning access inequities and the ongoing necessity for targeted, age-specific policies to address diverse weather impacts. Intergenerational Self-Help

Clubs provide vital community-based support, enhancing resilience against various climate shocks, including those from extreme weather events (Dao et al., 2020; Giang, 2013).

In **Nepal**, a country highly susceptible to a range of climate-related impacts including floods, landslides precipitated by heavy rainfall, and heatwaves, older individuals are recognized as particularly vulnerable (Chalise, 2006; Khanal & Dangal, 2022; Readiness to Implement the Action Plan to Strengthen Regional Cooperation on Social Protection: Nepal, 2025; WHO, 2016). Diminished mobility significantly impedes the capacity of older persons to evacuate swiftly from rapidly rising floodwaters in the Terai plains or regions vulnerable to landslides during intense monsoon rains in hilly areas ((Dixit, 2003; Khanal & Dangal, 2022). Furthermore, flood events substantially elevate the incidence of waterborne diseases and food insecurity among older adults (Khanal & Dangal, 2022; Readiness to Implement the Action Plan to Strengthen Regional Cooperation on Social Protection: Nepal, 2025). Nepal's Social Security Allowance (SSA) program provides coverage to multiple vulnerable demographics, encompassing its elderly population (UNICEF, 2019). The Nepal Red Cross Society has implemented pilot programs for shock-responsive social protection interventions, leveraging the existing SSA system to disseminate early warnings before anticipated flood events and to deliver timely financial assistance for recovery, thereby demonstrating the effective utilisation of established frameworks for disaster response related to weather events (Red Cross EU Office, n.d.). Notwithstanding constitutional provisions for social security, the provision of targeted age-appropriate services and climate-resilient health infrastructure, which are imperative during and after extreme weather events, remains constrained (Gentle & Maraseni, 2012; Khanal & Dangal, 2022; South Asian Voices, 2024).

Myanmar exhibits a swiftly growing elderly demographic that is markedly vulnerable to the effects of climate change. These effects include cyclones, significant fluctuations in precipitation patterns leading to both inundations and droughts, and escalating ambient temperatures. A considerable fraction of the aged population with disabilities experiences heightened susceptibility to weather-related catastrophes such as cyclones and floods. Mobility restrictions and pre-existing medical conditions increase risks during severe meteorological events, and older persons may be unintentionally neglected during rescue efforts related to floods or storms. The catastrophic consequences of Cyclone Nargis on the elderly were substantial, precipitating enduring obstacles to livelihood security and post-disaster recovery (Zaw, 2017). Persistent internal strife further exacerbates humanitarian needs for older individuals impacted by the simultaneous effects of climatic shocks and displacement. Elevated levels of insufficient food access among this cohort are intensified by widespread environmental instability. Accessibility to healthcare and consistent revenue streams—both indispensable for managing weather effects—present substantial challenges. Myanmar's social protection infrastructure is in an evolutionary phase, featuring a social pension scheme and the establishment of Elderly Self-Help Associations. However, a prominent deficiency in age-disaggregated data concerning weather repercussions frequently leads to the exclusion of older persons from humanitarian response strategies and assistance allocation during cyclones or floods (Horton et al., 2017).

Pakistan is highly vulnerable to climate-induced disasters, including frequent and intense floods, such as those experienced in 2010 and 2022, and heatwaves. These events place significant strain on the healthcare system, leading to preventable deaths among older adults. Flooding exacerbates disease risks, particularly waterborne and respiratory illnesses, posing a substantial threat to the elderly. Extreme heat presents a growing danger, especially for older adults with underlying health conditions. Inadequate evacuation procedures before floods and insufficient living conditions during displacement contribute to illness and mortality (Kreutzmann & Schütte, 2011; Saleem, 2016. Many older individuals are forced to work in hazardous, heat-exposed conditions due to limited social assistance programs. Pakistan has implemented cash transfers post-flood through initiatives like the Citizens' Damage Compensation Program (CDCP) and the Benazir Income Support Programme (BISP). However, fragmentation and operational silos between social protection and disaster management agencies hinder coordinated, age-inclusive responses. A lack of data on climate impacts makes the suffering of older people largely invisible within official statistics nd planning (Arsad et al., 2022; Prina et al., 2024).

2.3 Social Protection: A Pillar of Climate Resilience

The preceding analysis of vulnerabilities and prevailing national responses unequivocally demonstrates the imperative for social protection systems to evolve from a largely reactive approach to an essential, proactive component within comprehensive climate resilience strategies for elderly populations throughout Asia. These strategies must be specifically adapted to the challenges presented by extreme weather events. Adaptive Social Protection (ASP), by its intrinsic nature, amalgamates fundamental aspects of conventional social protection, disaster risk reduction (with a particular focus on weather-related perils), and climate change adaptation (Kuriakose et al., 2012). This amalgamation is intended to develop resilient systems capable of efficiently providing support to susceptible demographics prior to, during, and subsequent to the occurrence of meteorological shocks (GIZ, 2024; Ulrichs & Slater, 2016; UNICEF, 2022).

Adaptive Social Protection (ASP) frameworks facilitate critical anticipatory measures against predicted extreme weather occurrences. By strategically linking social protection beneficiary databases with meteorological early warning systems, support mechanisms can be established and activated based on reliable forecasts. This allows for proactive interventions, such as the prompt dissemination of early evacuation warnings for imminent floods or cyclones, the strategic pre-positioning of essential relief supplies in areas projected to be impacted by storms, or the provision of advance financial assistance before the commencement of a significant heatwave. This methodology effectively utilises existing social protection infrastructure for expedited mobilisation and response. ASP augments the shock responsiveness of safety nets, particularly in instances of weather-induced repercussions. Mechanisms that include pre-approved beneficiary rosters and surge capacity funding, potentially bolstered by integration with Disaster Risk Financing and Insurance (DRFI) instruments tailored for climatic events, enable the rapid vertical or horizontal expansion of assistance. This scalability is paramount during or immediately subsequent to a major flood, typhoon, or widespread drought, to address acute requirements for sustenance, refuge, medical care, and fundamental necessities that may have been compromised or rendered inaccessible due to the severity of the meteorological event. The Emergency Cash Transfer program in the Philippines exemplifies this strategic transition towards the expeditious provision of monetary aid following a weather-related crisis.

In addition to offering immediate assistance following extreme weather incidents, Adaptive Social Protection (ASP) plays a crucial role in developing long-term adaptive capacity and resilience to future climate-related events. Consistent social protection provisions can facilitate livelihood diversification, encouraging transitions from sectors highly susceptible to weather variability (e.g., rain-dependent agriculture vulnerable to drought or flooding). Furthermore, such provisions can enable investments in weather-resistant infrastructure and climate-resilient agricultural methods. Post-disaster support, provided subsequent to events such as floods or storms, can expedite the recovery of livelihoods and assets, preventing asset erosion or the onset of chronic poverty, often resulting from severe weather disruptions. Notably, for older individuals, ASP initiatives must deliberately incorporate age-specific components and actively address prevailing institutional and socio-cultural obstacles that hinder effective weather preparedness and response for this demographic. This requires not only customising aid to address needs exacerbated by weather conditions (e.g., ensuring uninterrupted access to healthcare and clean water during floods) but also promoting the genuine inclusion of older persons in community resilience planning. Such engagement enables the utilisation of their significant local knowledge of historical weather phenomena and conventional adaptation strategies. Initiatives that foster intergenerational collaboration further illustrate the beneficial outcomes of this inclusive strategy in enhancing community resilience against climatic shocks.

Transforming the significant potential of Adaptive Social Protection (ASP) into concrete, beneficial results for older individuals exposed to extreme weather necessitates a focused endeavour to confront enduring challenges observed throughout the Asian region. These challenges encompass the fragmentation of social protection and disaster management frameworks (Devereux & Sabates-Wheeler, 2004), inadequate earmarked funding for climate resilience strategies tailored to specific meteorological hazards, an absence of detailed, age-disaggregated data on weather repercussions to facilitate accurate targeting and robust monitoring (Waha et al., 2017), and the widespread disparity between national policy directives and local implementation systems for weather-related assistance (HelpAge USA, 2025; Oxford Policy Management, 2017; Razzaque et al., 2025). Enhancing administrative capabilities, particularly at devolved tiers of governance, is of critical importance for guaranteeing precise identification of vulnerable older individuals and efficacious provision of aid during and in the aftermath of weather-induced emergencies (Agrawal, 2008).

2.4 Conclusion and Implications

This chapter synthesises the intricate and convergent crises of rapid demographic aging and intensifying extreme weather events in Asia, presenting an analysis of the multifaceted vulnerabilities experienced by older populations. The reviewed evidence demonstrates that vulnerability to hydro-meteorological events, such as heatwaves, floods, cyclones, and droughts, is not solely attributable to advanced age. Instead, it stems from complex and dynamic interactions between physiological susceptibilities, exacerbated by thermal stress or other weather-related distress (Patz et al., 2005), entrenched socio-economic disparities (Chambers, 1995), and critical inadequacies in existing institutional and policy frameworks concerning weather-related disaster risk reduction and social protection (Brooks et al., 2005). Across a range of national contexts—including the direct health consequences of heatwaves and the significant challenges posed by floods and cyclones in China and India, disaster-related mortality linked to floods, heatwaves, and typhoons in Japan and the Philippines, the severe impacts of floods and storms in Bangladesh, Indonesia, and Thailand, the compounded weather risks in Vietnam, Nepal, and Pakistan, and the devastating consequences of cyclones in Myanmar—older persons consistently endure a disproportionate burden of climate change impacts specifically related to extreme weather events (Bartlett, 2008).

This analysis demonstrates the essential role of adaptive social protection (ASP) as a proactive strategy to address the nexus of extreme weather and vulnerability among older persons. ASP frameworks, designed for anticipatory action using weather forecasting, responsive measures to immediate shocks, and sustained support for climate-induced losses, are crucial for enhancing the resilience of older populations. Nevertheless, realising this potential necessitates addressing systemic challenges. These include policy and operational disconnects between social protection and disaster management entities, a lack of age-specific considerations in response plans, a limited evidence base due to insufficient disaggregated data, persistent resource limitations for targeted interventions, and exclusionary practices that hinder the participation of older persons in planning (McCracken & Phillips, 2016).

From this synthesis and critical analysis, several substantial scholarly implications and imperatives for policy and practice are evident regarding the well-being of older persons in extreme weather conditions. Initially, there is an explicit exigency for integrated research initiatives that surpass customary disciplinary boundaries. This research must concurrently investigate the intricate interrelation between gerontology (Settersten Jr & Angel, 2011), climate science (concentrating particularly on extreme weather phenomena), disaster risk reduction (focused on weather-related perils), and social policy (Spicker et al., 2008). A fundamental priority for this research agenda must be the collection and meticulous analysis of disaggregated data delineating the impacts of weather events by age, gender, disability status, and other relevant indices of vulnerability. This will yield compelling evidence crucial for the formulation and execution of precise anticipatory and responsive interventions (Amnesty International, 2025; HelpAge USA, 2025; Moser & Satterthwaite, 2008; UNFPA Vietnam, 2024).

A critical evaluation of existing social protection systems, viewed through a climate vulnerability lens, is imperative within the scope of social policy research. This evaluation must determine specific mechanisms to enhance responsiveness to shocks and ensure the inclusivity of older adults. Consideration of the diverse weather hazards prevalent in each country is paramount (Gough, 2004; Oxford Policy Management, 2018; Razzaque et al., 2025; UNICEF, 2022). Research endeavours should investigate methods to adapt or augment current systems, thereby addressing the nuanced needs of older individuals prior to, during, and subsequent to extreme weather events.

It is imperative to conduct empirical research into the effectiveness of age-inclusive Disaster Risk Reduction (DRR) and Climate Change Adaptation (CCA) strategies pertinent to weather events (Van Aalst et al., 2008). Specifically, such research should scrutinise the challenges and accomplishments associated with translating national policy objectives into tangible local-level actions. Critical areas requiring investigation encompass the efficacy of early warning systems for heatwaves, floods, and storms in reaching older populations, the suitability of evacuation procedures for individuals with mobility or health impairments, and the accessibility and sufficiency of post-disaster support services, while also addressing administrative impediments and attitudinal obstacles that may impede access for older adults (Few, 2003; Guanio, 2021; Tamura et al., 2009).

In conclusion, it is imperative to conduct research into innovative and contextually appropriate financing mechanisms for climate-resilient social protection (Mechler et al., 2009). This is essential to guarantee the long-term sustainability and scalability of interventions required to mitigate the impacts of increasingly frequent and intense extreme weather events (CEEW, 2025; UNICEF Indonesia, 2024; World Bank, 2024). Investigating alternatives such as dedicated climate funds, insurance schemes, and budget reallocation strategies will be of critical importance. A holistic, proactive, and rights-based approach is imperative to address the escalating nexus of aging and climate change in Asia, particularly the heightened risks from extreme weather, which explicitly centres the unique needs, vulnerabilities, and capabilities of older persons (UNDP, 2007). By fortifying adaptive social protection systems and methodically integrating aging considerations into climate and weather-related disaster resilience strategies, Asian countries can advance towards a future characterised by substantially reduced vulnerability to intensifying shocks. Such actions will ensure the dignity, security, and well-being of all citizens, including the region's rapidly growing older population, amidst increasing climatic instability. Future research endeavors must meticulously investigate the efficacy of specific Adaptive Social Protection (ASP) interventions across diverse climatic contexts and for distinct sub-groups within the older population (e.g., rural versus urban dwellers, individuals with varying types and degrees of disability) to continuously refine and optimise adaptive responses to the region's multifaceted weather hazards (Smit & Wandel, 2006).

References

Adger, W. N. (2006). Vulnerability: A generally applicable conceptual framework for climate change research. *Global Environmental Change, 16*(3), 268–281.

Afrin, S., Khan, M. M. H., & Haque, M. A. (2025). Factors affecting the active aging situation in Bangladesh. *Frontiers in Public Health, 13*, 1517482.

Agrawal, A. (2008). *The role of local institutions in adaptation to climate change.* Social Development Papers, Paper No. 118. World Bank.

Almazan, J. U., Albougami, A. S., Alamri, M. S., Colet, P. C., Adolfo, C. S., Allen, K., et al. (2019). Disaster-related resiliency theory among older adults who survived Typhoon Haiyan. *International Journal of Disaster Risk Reduction, 35*, 101070.

Amnesty International. (2025, May 5). *Pakistan: Uncounted: Invisible deaths of older people and children during climate disasters in Pakistan.* https://www.amnesty.org/en/documents/asa33/9007/2025/en/

Arsad, F. S., Hod, R., Ahmad, N., Ismail, R., Mohamed, N., Baharom, M., et al. (2022). The impact of heatwaves on mortality and morbidity and the associated vulnerability factors: A systematic review. *International Journal of Environmental Research and Public Health, 19*(23), 16356.

Asian Development Bank. (2025). Adapting to aging Asia and the Pacific. https://www.adb.org/what-we-do/topics/social-development/aging-asia

Bankoff, G. (2003). Constructing vulnerability: The historical, natural and social generation of flooding in metropolitan Manila. *Disasters, 27*(3), 224–238.

Barrientos, A., & Hulme, D. (2008). *Social protection for the poor and poorest: Concepts, policies and politics.* Palgrave Macmillan.

Bartlett, S. (2008). Climate change and urban children: Impacts and implications for adaptation in low- and middle-income countries. *Environment and Urbanization, 20*(2), 501–519.

Brooks, N., Adger, W. N., & Kelly, P. M. (2005). The determinants of vulnerability and adaptive capacity at the national level and the implications for adaptation. *Global Environmental Change, 15*(2), 151–163.

Carter, P., Gyamfi, C. O., Quddus, T., Ghandali, M. V., & Hawken, E. (2025). Comparing the Impact of Flooding on Mental Health in India and the United Kingdom: Who Is More Vulnerable? *BJPsych Open, 11*(S1), S27–S27.

Chalise, H. N. (2006). Social support and its correlation to loneliness and depression among older adults in Nepal. *Kathmandu University Medical Journal, 4*(1), 16–20.

Chambers, R. (1995). Poverty and livelihoods: Whose reality counts? *Environment and Urbanization, 7*(1), 173–204.

Chatterjee, A., Wadhawan, S., & Chitale, V. (2025). Impact of heat exposure on health outcomes among older adults in India: An analysis across ten states. *Ageing & Society, 1*, 1–22.

China CDC. (2025, February 21). *China national climate change health adaptation action plan (2024–2030).* https://en.chinacdc.cn/health_topics/environment_health/202502/t20250221_304471.html

Council on Energy, Environment and Water (CEEW). (2025, May 20). *How extreme heat is impacting India: Assessing district-level heat risk.* https://www.ceew.in/publications/mapping-climate-risks-and-impacts-of-extreme-heatwave-disaster-in-indian-districts

Dao, B. T., Tran, N. T., Barysheva, G. A., & Tran, L. S. (2020). Social security and population ageing in Vietnam: A guarantee for the elderly people's life. *International Journal of Criminology and Sociology, 9*, 381–390.

Devereux, S., & Sabates-Wheeler, R. (2004). *Transformative social protection* (IDS Working Paper, October).

Dixit, A. (2003). Floods and vulnerability: Need to rethink flood management in Nepal. In M. M. Ballabh, S. L. Vaidya, & C. S. Kumar (Eds.), *Water resources development in Nepal* (pp. 211–230). Books Faith India.

DSWD. (2021). *DSWD Guidelines on Emergency Cash Transfer*. DSWD.

ESCAP. (2025, February). *Readiness to implement the action plan to strengthen regional cooperation on social protection: Nepal*. ESCAP.

ESCAP (Economic and Social Commission for Asia and the Pacific). (2017). *Ageing in Asia and the Pacific: Overview*. United Nations.

Few, R. (2003). Flooding, vulnerability and coping strategies: Local responses to a global threat. *Progress in Development Studies, 3*(1), 43–58.

Gentle, P., & Maraseni, T. N. (2012). Climate change, poverty and livelihoods: Adaptation practices by rural mountain communities in Nepal. *Environmental Science & Policy, 21*, 24–34.

Giang, T. L. (2013). Social protection for older people in Vietnam: Role, challenges and reform options. *Malaysian Journal of Economic Studies, 50*(2), 207–219.

GIZ. (2024). *Advancing adaptive social protection in Indonesia*. https://www.giz.de/en/worldwide/155135.html

GlobalGiving. (2011, December 1). *Help disabled and elderly disaster survivors*. https://www.globalgiving.org/projects/help-disabled-elderly-and-displaced-japanese/reports/?subid=17109

Gough, I. (2004). Human well-being and social structures: Relating the universal and the local. *Global Social Policy, 4*(3), 289–311.

Guanio, K. R. G. (2021). Disasters are Ageist, is DRRM too? *Philippine Population Review, 60*(1), 111–125.

Hasan, M. M. (2012). Effectiveness of old age allowance program in Bangladesh: A study on the demand side dynamics. *Master in Public Policy and Governance Program, Department of General and Continuing Education, North South University, Bangladesh.*

Hashim, J. H., & Hashim, Z. (2016). Climate change, extreme weather events, and human health implications in the Asia Pacific region. *Asia Pacific Journal of Public Health, 28*(2_suppl), 8S–14S.

Heijmans, A. (2004). From vulnerability to empowerment: A paradigm for development. In G. Bankoff, G. Frerks, & D. Hilhorst (Eds.), *Mapping vulnerability: Disasters, development and people* (pp. 117–129). Earthscan.

HelpAge International. (2008). The situation of older people in cyclone-affected Myanmar: Nine months after the disaster.

HelpAge International. (2020). Income security for all older people in Bangladesh during COVID-19 and beyond.

HelpAge International. (2022, November 28). *Older citizen monitoring: A global review*. https://www.ageinternational.org.uk/siteassets/documents/ocm_report_final_web.pdf

HelpAge International. (2024, December). *Ensuring inclusive climate action in Bangladesh*. https://www.helpage.org/wp-content/uploads/2024/12/Ensuring-inclusive-climate-action-in-Bangladesh.pdf

HelpAge USA. (2025, March 31). *Older people must be central to Myanmar's earthquake relief efforts*. https://helpageusa.org/older-people-must-be-central-to-myanmars-earthquake-relief-efforts/

Heltberg, R., Siegel, P. B., & Jorgensen, S. L. (2009). Addressing human vulnerability to climate change: Toward a 'no-regrets' approach. *Global Environmental Change, 19*(1), 89–99.

Horton, R., De Mel, M., Peters, D., Lesk, C., Bartlett, R., Helsingen, H., et al. (2017). *Assessing climate risk in Myanmar: Technical report*. Center for Climate Systems Research at Columbia University, WWF-US and WWF-Myanmar.

Huq, S., Karim, Z., Asaduzzaman, M., & Mahtab, F. (1999). *Vulnerability and adaptation to climate change for Bangladesh*. Kluwer Academic Publishers.

Islam, M. R., & Rashid, M. M. (2011). Riverbank erosion in Bangladesh: A case study of two upazilas in Sirajganj district. *Journal of Geography and Regional Planning, 4*(6), 327–338.

Japan Centre for International Exchange. (2024, February). *AHWIN paper no. 4: Japan's experience in strengthening disaster management and relief measures for older people*. http://jcie.org/wp-content/uploads/2024/02/AHWIN-Paper-no4-0224.pdf

Jha, R. (2014). Welfare schemes and social protection in India. *International Journal of Sociology and Social Policy, 34*(3/4), 214–231.

Juliannisa, I. A., Rahma, H., Mulatsih, S., & Fauzi, A. (2025). Regional vulnerability to food insecurity: The case of Indonesia. *Sustainability, 17*(11), 4800.

Khan, S., & Marzi, M. (2019, May). *Strengthening the capacity of ASEAN Member States to design and implement risk-informed and shock-responsive social protection systems for resilience.*

Khanal, D., & Dangal, M. R. (2022). Environmental social work: Mitigating the impact of environmental crises on older people living in disaster prone areas in Nepal. *Bodhi: An Interdisciplinary Journal, 8*(8), 1–16.

Kinsella, K., & Phillips, D. R. (2005). Global aging: The challenge of success. *Population Bulletin, 60*(1), 1–40.

Kreutzmann, H., & Schütte, S. (2011). The 2010 flood in Pakistan: A review of the hydrometeorological and geopolitical dimensions. *Geographische Rundschau International Edition, 7*(1), 4–11.

Kuriakose, A. T., Heltberg, R., Wiseman, W., Costella, C., CIPR, C. S., & Mahrt, K. (2012). *Climate-responsive social protection.* World Bank.

Lam, Y. F., & Roy, S. (2019). Climate adaptation of sea-level rise in Hong Kong. In R. Akhtar (Ed.), *Extreme weather events and human health: International Case Studies* (pp. 117–130). Springer International Publishing. https://doi.org/10.1007/978-3-030-23773-8

Lutz, W., Sanderson, W., & Scherbov, S. (2001). The end of world population growth. *Nature, 412*(6846), 543–545.

Mazdiyasni, O., Agha Kouchak, A., Davis, S. J., Madadgar, S., Mehran, A., Ragno, E., et al. (2017). Increasing probability of record-shattering regional daily temperatures. *Science Advances, 3*(7), e1700044.

McCracken, K., & Phillips, D. R. (2016). Climate change and the health of older people in Southeast Asia. In R. Akhtar (Ed.), *Climate change and human health scenario in South and Southeast Asia* (pp. 29–52). Springer International Publishing. https://doi.org/10.1007/978-3-319-23684-1_3

Mechler, R., Linnerooth-Bayer, J., & Peppiatt, D. (2009). *Disaster insurance for the poor? A review of microinsurance for natural disaster risks in developing countries.* ProVention Consortium.

Meehl, G. A., & Tebaldi, C. (2004). More intense, more frequent, and longer lasting heat waves in the 21st century. *Science, 305*(5686), 994–997.

Ministry of Social Justice and Empowerment. (2020). *An umbrella scheme for senior citizens.* Government of India. https://grants-msje.gov.in/display-napsrc

Mohanty, C. R., Jain, M., Radhakrishnan, R. V., & Mohanty, P. C. (2020). Tropical cyclone Fani–perspective from the trauma and emergency department of an affected tertiary hospital. *Chinese Journal of Traumatology, 23*(4), 243–248.

Moser, C., & Satterthwaite, D. (2008). *Towards pro-poor adaptation to climate change in the urban centres of low and middle-income countries.* International Institute for Environment and Development (IIED).

Nadiruzzaman, M. D. (2012). *Cyclone Sidr and its aftermath: everyday life, power and marginality* (Doctoral dissertation, Durham University).

Nakai, S., Itoh, T., & Morimoto, T. (1999). Deaths from heat-stroke in Japan: 1968-1994. *International Journal of Biometeorology, 43*(3), 124–127.

National Economic and Development Authority. (2017). *Philippine development plan 2017-2022.* https://pdp.neda.gov.ph/wp-content/uploads/2017/01/11-04-07-2017.pdf

O'Brien, K., Leichenko, R., Kelkar, U., Venema, H., Aandahl, G., Tompkins, H., et al. (2006). Mapping vulnerability to multiple stressors: Climate change and globalization in India. *Global Environmental Change, 16*(2), 196–213.

Oxford Policy Management. (2017). *Shock-responsive social protection systems: Policy brief, Pakistan.* Oxford Policy Management.

Oxford Policy Management. (2018, December). *ASEAN report: Thailand SRSP case study final.* Oxford Policy Management.

Partridge, L., Deelen, J., & Slagboom, P. E. (2018). Facing up to the global challenges of ageing. *Nature, 561*(7721), 45–56.

Patz, J. A., Campbell-Lendrum, D., Holloway, T., & Foley, J. A. (2005). Impact of regional climate change on human health. *Nature, 438*(7066), 310–317.

Poapongsakorn, N., & Meethom, P. (2013). Impact of the 2011 floods, and flood management in Thailand. *ERIA Discussion Paper Series, 34*, 2013.

Prina, M., Khan, N., Khan, S. A., Caicedo, J. C., Peycheva, A., Seo, V., et al. (2024). Climate change and healthy ageing: An assessment of the impact of climate hazards on older people. *Journal of Global Health, 14*, 04101.

Raghavan, S., & Rajesh, S. (2003). Trends in tropical cyclone impact: A study in Andhra Pradesh, India. *Bulletin of the American Meteorological Society, 84*(5), 635–644.

Razzaque, M. A., Rahman, M. M., & Hossain, M. A. (2025, March). *Strengthening social protection in Bangladesh: What reforms are most urgently needed?* International Growth Centre.

Red Cross EU Office. (n.d.). *Piloting shock responsive social protection in Nepal through cash assistance.* https://redcross.eu/projects/piloting-shock-responsive-social-protection-in-nepal-through-cash-assistance

Refworld. (2024, December 6). *Philippines: Republic Act No. 12076 (Ligtas Pinoy Centers Act).* https://www.refworld.org/legal/legislation/natlegbod/2024/en/149347

Reyes, C. M., Albert, J. R. G., & Reyes, C. C. M. (2018). *Lessons on providing cash transfers to disaster victims: A case study of UNICEF's unconditional cash transfer program for super typhoon Yolanda victims* (No. 2018–04). PIDS Discussion Paper Series.

Roy, K., & Hulme, D. (2003). *Chronic poverty in Bangladesh: tales of ascent, descent, stagnation and resilience.* Chronic Poverty Research Centre.

Saleem, S. (2016). Pakistan floods: Reports from the field. In L. Racioppi & S. Rajagopalan (Eds.), *Women and Disasters in South Asia: Survival, security and development* (pp. 192–207). Routledge India. https://www.taylorfrancis.com/books/edit/10.4324/9781315650630/women-disasters-south-asia-linda-racioppi-swarna-rajagopalan

Settersten, R. A., Jr., & Angel, J. L. (2011). *Handbook of sociology of aging.* Springer.

Shaw, R., & Takeuchi, Y. (2005). *Indigenous knowledge and disaster risk reduction: From practice to policy.* Nova Science Publishers.

Smit, B., & Wandel, J. (2006). Adaptation, adaptive capacity and vulnerability. *Global Environmental Change, 16*(3), 282–292.

South Asian Voices. (2024, May 15). *Climate and health in Nepal: A federalism perspective.* https://southasianvoices.org/ec-m-oth-n-climate-health-nepal/

Spicker, P., Leguizamon, S. A., & Gordon, D. (2008). *Poverty: An international glossary* (2nd ed.). Zed Books.

Stockholm Environment Institute. (2024). *Global ageing and climate change: Asia and the Pacific webinar summary.* https://www.sei.org/wp-content/uploads/2024/04/global-ageing-climate-change-asia-and-the-pacific-webinar-summary-01072024.pdf

Sun, S., Tian, L., Qiu, H., Chan, K. P., Tsang, H., Tang, R., et al. (2016). The influence of pre-existing health conditions on short-term mortality risks of temperature: Evidence from a prospective Chinese elderly cohort in Hong Kong. *Environmental Research, 148*, 7–14.

Tamura, K., Hayashi, H., & Kimura, R. (2009). Clarifying the disaster process of the elderly in the aspect of social welfare. *Disaster Prevention Research Institute Annuals, 52*(B), 16–25.

The Japan Times. (2018). *Protect the socially weak from heat waves.* The Japan Times. https://www.japantimes.co.jp/opinion/2018/07/25/editorials/protect-socially-weak-heat-waves/.

The Judiciary PH. (2024, December 10). *Republic act no. 12076-an act establishing evacuation centers for every city and municipality, and appropriating funds therefor.* https://elibrary.judiciary.gov.ph/thebookshelf/showdocs/2/98122

Tian, Q., & Lemos, M. C. (2018). Household livelihood differentiation and vulnerability to climate hazards in rural China. *World Development, 108*, 321–331.

Ulrichs, M., & Slater, R. (2016). *How can social protection build resilience? Insights from Ethiopia, Kenya and Uganda* (BRACED Working Paper). ODI.

UNDP (United Nations Development Programme). (2007). *Fighting climate change: Human solidarity in a divided world. Human Development Report 2007/2008.* Palgrave Macmillan.

UNDRR, United Nations Office for Disaster Risk Reduction. (2024). *GAR Special Report 2024.*

UNFPA Vietnam. (2024). *Policy brief: Inclusion of older people in disaster risk reduction.* https://vietnam.unfpa.org/sites/default/files/resource-pdf/policy_brief_-_inclusion_of_op_in_drr_unfpa_eng_final.pdf

UNICEF. (2019, June). International social protection conference: Resilient social protection for an inclusive future.

UNICEF. (2022, March 15). *Strengthening Viet Nam's social protection system with a focus on child-sensitivity and shock-responsiveness.*

UNICEF Indonesia. (2024, March). *Integrating adaptive social protection financing into disaster risk financing strategy.*

Van Aalst, M. K., Cannon, T., & Burton, I. (2008). Community level adaptation to climate change: The potential role of participatory community risk assessment. *Global Environmental Change, 18*(1), 165–179.

Waha, K., Krummenauer, L., Adams, S., Aich, V., Baarsch, F., Coumou, D., et al. (2017). Climate change impacts in the Middle East and Northern Africa (MENA) region and their implications for vulnerable population groups. *Regional Environmental Change, 17*(6), 1623–1638.

Wisner, B., Blaikie, P., Cannon, T., & Davis, I. (2004). *At risk: Natural hazards, people's vulnerability and disasters* (2nd ed.). Routledge. https://www.routledge.com/At-Risk-Natural-Hazards-Peoples-Vulnerability-and-Disasters/Blaikie-Cannon-Davis-Wisner/p/book/9780415252157

World Bank. (2024, May 28). *How intergenerational self-help clubs transform elderly care in Viet Nam.*

World Health Organisation (WHO). (2016). *Climate change and health in Nepal: A country profile.* WHO.

Yang, J., & Hayashi, H. (2023). Increased care needs in older long-term care insurance users after the 2018 Japan Floods. *BMC Geriatrics, 23*(1), 1–9.

Yang, Y., & Zhang, Y. (2025). Climate risks, multi-tier medical insurance systems, and health inequality: Evidence from China's middle-aged and elderly populations. *BMC Public Health, 25*(1), 1–13.

Yao, Y., Li, T., Li, Y., & Zhang, Y. (2025). Elderly vulnerability to temperature-related mortality risks in China. *Science Advances, 11*(5), eado5499.

Yoshida, S., Kashima, S., Ishii, S., Koike, S., & Matsumoto, M. (2022). Effects of the 2018 Japan floods on long-term care insurance costs in Japan: retrospective cohort study. *BMC Public Health, 22*(1), 341.

Zaw, M. (2017). *Re-visiting cyclone Nargis of Myanmar: Vulnerability, resiliency and the dynamics of recovery in the Ayeyarwady Delta.* Australian National University.

Chapter 3
Vulnerability to Extreme Weather Events

The concept of vulnerability reflects the fact that we all are born, live, and die within a fragile materiality that renders all of us constantly susceptible to destructive external forces and internal disintegration. (Martha Albertson Fineman, 2012)

3.1 The Concept of Vulnerability

The concept of vulnerability has its origins in Latin words, the noun vulnus ('to wound'), the verb vulnerare ('to wound') and the Late Latin adjective vulnerabilis, which was adopted into English in the form of 'vulnerable' in the early 1960s (Forbes-Mewett & Nguyen-Trung, 2019; Ford et al., 2018; Timmerman, 1981). In the 1970s, the concept was predominantly used in the natural and environmental hazards studies to understand the consequences of natural disasters and coping measures (Burton et al., 1978, 1993; Wisner, 2016; Wisner & Luce, 1993). However, Kelman et al. (2016) argued, while developing the concepts of vulnerability and resilience, that there was a formal overlap of literature on disaster research with ongoing work in international development (for details, Ball, 1975; Burton et al., 1978; O'Keefe, 1976; Timmerman, 1981).

Since the late 1980s, societies have been increasingly exposed to risks and hazards that add a new dimension to the concepts of vulnerability and adaptive capacity to climate change and extreme weather events, even as development studies have evolved to explore the impacts of poverty and progress on people and society (Chambers, 1989; Liverman, 1990; Siegel & Alwang, 1999). The increasing risks of natural disasters, technological failure, terrorist attacks, nuclear wars and environmental degradation suggest that people are becoming more vulnerable to extreme events (Forbes-Mewett & Nguyen-Trung, 2019). Thus, from the beginning of the twenty-first century the terms 'vulnerability', 'resilience' and 'adaptive capacity' are frequently used, especially when studying disasters through a development lens

M. I. Ali et al., *Extreme Weather Events and Vulnerability of Older People*,
https://doi.org/10.1007/978-3-032-11256-9_3

(Adger, 2006; Kelman et al., 2016; IPCC, 2001; IPCC, 2001, 2014; UNDP, 2004; UNISDR, 2015).

However, opinions on who is vulnerable and why varies widely across research fields and over time. Children and young people, older people, and ethnic minorities are often considered vulnerable (Virokannas et al., 2020). Some researchers believe that vulnerability should not be considered for a specific person or group but as a universal human condition (Fineman, 2008, 2016).

The definition of the vulnerability of human beings to extreme events varies across disciplines over time and context. Timmerman (1981) defined vulnerability as "the degree to which a system responds adversely to the occurrence of a hazardous event. The degree and quality of the adverse response depend on a system's resilience (a measure of the system's capacity to absorb and recover from the event)." Chambers (1989, p. 1) defines vulnerability as "exposure to contingencies and stress, and difficulty in coping with them." This definition has an external aspect of hazards and shocks to which individuals or households are exposed, as well as an internal component, namely the lack of resources to cope with the risks, threats or hazards (Crooks, 2009). According to Clark et al. (1998), vulnerability is a product of two properties: exposure (the probability of experiencing a hazardous event) and coping capacity, which can be further subdivided into resilience and resistance (the ability to recover from losses following the impact of an extreme weather event). It refers to vulnerability as a result of intense exposure to hazards and uncertainties combined with limited capacity to protect oneself from these hazards or to cope with their adverse effects (United Nations, 2003). The vulnerability has dynamic properties and multiple stresses that accelerate threats, reduce coping capacities and bring changes in risk exposure.

Further, vulnerability can be defined as the conditions of a group or community determined by physical, social, economic and environmental factors or processes in the places where they live that increase their susceptibility and predisposition to the effects of hazards following an extreme weather event (Cardona, 2006; Van Westen, 2014). Indeed, the presence of vulnerability in the environment or context depends on the areas in which people live and how their settlements, social profile of communities and livelihoods are structured. The potential for damage and the complex interaction between risk, mitigation and the social network of a place constitute the social vulnerability to extreme natural hazards. From the above, it is clear that vulnerability is a complex construct encompassing various elements of environment, settlement, society, culture, economy and government, and its standard or consensus definition is a real challenge due to its components, dynamics and complexity. This is because it results not only from external sources (e.g. extreme weather events) but also from uncertain conditions deeply rooted in social, political and economic systems.

Thus, the vulnerability of older people to extreme weather events can be viewed as a cause-effect relationship between extreme weather events and their impact on older people and their households, their economic conditions and socio-ecological systems (Fritzsche et al., 2014). For this study, the vulnerability of older people is defined as a condition resulting from the consequences of the cumulative impact of extreme weather events combined with a lack of capacity to access socio-economic resources and social protection programmes.

3.2 The Vulnerability of Older People

Human beings are inherently vulnerable when they do not have the capabilities to exercise choice, justice and freedom, capital assets and support systems to deal with threats they face (Sen, 1980). The vulnerability may begin from infancy or at any stage of life and influence the rest of the lifespan. Household poverty, livelihood insecurity, and lack of education and employment adversely affect building human ability and socioeconomic capabilities that affect economic achievement and social functioning and influence vulnerability (Emmerson & Muriel, 2008; McLaughlin & Jensen, 2000; Zaidi, 2014). Moreover, socioeconomic discrimination in earlier life, depending on socioeconomic class, religion, ethnicity, gender, caste, age, employment and income may influence the vulnerability of people in later life and even generations. According to the linked lives principle of the life course approach, a person's life trajectory is linked to the life trajectories of the people they are connected to, especially among household members (Levy et al., 2016). Therefore, the vulnerability of human beings can be considered as the vicious circle that not only affects a person's life course but also transmits across the life domains of future generations. Thus, the vulnerability of older people is the result of processes that take place throughout their lives.

Furthermore, older people become vulnerable when there is an intergenerational transmission of poverty that displaces family members from the family network (due to displacement and migration) and prevents them from offering adequate assistance (Kreager, 2006). People with low-income parents are likely to persist low-income themselves. Similarly, people whose parents are wealthy are likely to remain rich as parents who have more resources and opportunities can pass on more resources and opportunities to their children (Rank, 2008). The frequent extreme weather events also impact the organisation of traditional family structure through family members' migration to other places for survival, which increases the socio-economic vulnerabilities among older people (Bates et al., 2012; Chartered Institute of Environmental Health, 2008). Besides, displacement and migration of family members are most likely to force older people to live alone at home during extreme weather events, increasing their risk of injury and death compared to others (Feist, 2016). The loss of family members, caregivers, and social networks can also lead to isolation in older people, which can force them to be vulnerable to extreme weather events.

The vulnerability of older people is a multidimensional construct that originates from interrelated sources and contexts — personal, socioeconomic, environmental, political and infrastructural deterioration and the failure of the governance system of the affected region (Cutter et al., 2003; Dwyer et al., 2004). The commonly recognised factors that influence social vulnerability include a lack of access to resources:

> social networks and connections, beliefs and customs, age, and the nature and density of infrastructure. (Blaikie et al., 2014; Tierney et al., 2001)

Again, Schröder-Butterfill and Marianti (2006) argued that the vulnerability of older people is the cumulative result of some different but interrelated risks, namely the risk of being exposed to a threat, the risk of a threat materialising, and the risk of not having the necessary defences to cope with a threat. Specifically, vulnerability in old age results from "the interplay between biological and social threats, individual characteristics and resources, social relationships and wider economic, political and cultural structures." Further, several personal characteristics (health, economic status, and social considerations) interact with contextual factors (the adequacy of emergency preparedness, transportation, and coping and recovery resources) to influence older adults' vulnerability to extreme weather events. This risk and threat of older people may be rooted in the social structure, connected to their access to socioeconomic, cultural, political and social resources, which cover a wide variety of interactions among families, peer groups and other social, cultural and political organisations. The interplay of these resources acts as the sources of developing coping capacity, or their deterioration can create vulnerability. Thus, the magnitude of the vulnerability of older people varies as they face challenges differently in the aftermath of extreme weather events due to their lack of resources, capacities and availability of support (Brooks et al., 2009; Heitzmann et al., 2002).

The vulnerability of older people is a lack of capacity deeply rooted in socioeconomic conditions, social structure, historical factors and the socio-governmental support system. Kelly and Adger (2000) describe social vulnerability as the capacity of an individual or group to respond to, cope with, recover from or adapt to external stresses (extreme events), focusing on socioeconomic and institutional restrictions that limit the ability to respond properly. They argue that a long-term solution must address the underlying causes of social vulnerability, such as the unequal distribution of resources. However, the capacity of older people to cope with the impacts of extreme weather events depends on a variety of personal factors, including wealth, education, skills and health, as well as social networks, including family, friends, neighbours and community institutions, and social protection programmes (Haq & Gutman, 2020; Van Eeuwijk, 2006). Again, chronic health problems, social isolation and limited access to resources and services, which are common among older people, can affect their socioeconomic capacity to cope with the consequences of extreme climatic events.

Grundy (2006), thus, argues that vulnerable older people are those who have a reserve capacity below the appropriate level to address the challenges (such as the consequences of extreme weather events) they face effectively. Every person reaches old age with a reserve, which has accumulated over the life course. This "reserve" includes income and material resources, family and social support and health status, which comprise old age security measures. If older people exceed their reserves and resources to minimise the challenges and risks or threats, they fall into vulnerability. Likewise, the economic capacity of older people to cope with the consequences of extreme events also depends on the availability and accessibility of assets and resources. The more resources people have, the less vulnerable they are. However, these resources are diverse and include human capital, psychological capital, social capital, information capital, economic capital, political capital and natural capital (Thomas et al., 2018). Thus, an individual or household may be described as vulnerable or vulnerable to an outcome (of extreme events) if they are incapable of dealing with the disaster with available resources (Alwang et al., 2001). Therefore, the vulnerability of older people in a socio-economic context is a complex and pre-existing condition that depends on both individual characteristics — age, gender, education, health and social and economic factors such as employment, income, access to credit, social connections and support.

3.3 Vulnerability to Extreme Weather Events

Extreme weather events are occurring with increasing frequency at many local, regional and global levels and have a devastating impact on socio-economic conditions, farms and human lives, food insecurity, and water scarcity (Obi et al., 2021; Raymond et al., 2020). Vulnerability to extreme weather events results from a complex interplay of climate events, environmental impacts and socio-economic dynamics in the affected society. The vulnerability depends on the significance, severity and duration of exposure, along with the characteristics of extreme weather events that make individuals and physical and social infrastructures vulnerable to the impacts (Oven et al., 2019). It may result in death, injury, and illness, as well as the destruction of property and the environmental ecosystem. All these losses lead to higher costs and prices, which vary by industry and region throughout the economy through trade relations and capital flows, leading to systemic consequences and structural changes in production, consumption and trade patterns (Calzadilla et al., 2006). However, the social impacts of extreme weather events occur through deteriorating social networks and relationships that could be more severe for marginalised groups such as older people (Curtis et al., 2017).

Various extreme weather events make older people vulnerable to the uncertainty of livelihoods, health and wellbeing, resettlement and food security in many countries, particularly developing and poor countries in Latin America, Africa and Asia (Otto et al., 2017). They argued that extreme weather events are likely to exacerbate current vulnerabilities and intergenerational inequalities. For example, early childhood deficits may limit future access to education and income opportunities later in life. Thus, the vulnerability of older people to extreme weather events generates adverse interactions between co-occurring factors such as poverty, social isolation, limited mobility, poor health and chronic illness, and inadequate or no health insurance, which influence their vulnerability to extreme weather events (Haq et al., 2008). Malak et al. (2020) showed that in northern coastal Bangladesh, older people were particularly vulnerable when a cyclone hit them due to their lack of assets, dependence on young adult family members and declining physical strength and mental abilities. Their vulnerability was particularly high because of a lack of social awareness, education and limited access to health facilities. Older people often lack adequate sanitation and hygiene facilities, food security and family care and social support from the government.

Extreme weather events have the potential to negatively affect the environment in many ways increasing vulnerability to everyone, but disproportionately affecting the most unprotected populations, such as older people and marginalised groups. Their livelihoods, the regions they live in and the rights and opportunity to access available assets (human, social, financial, physical and natural) influence the vulnerability of older people to extreme climatic events. Older people are particularly vulnerable to environmental impacts due to their age, limited mobility, changes in physiology and limited access to resources, which can affect their ability to adapt to environmental degradation following extreme climatic events (Filiberto et al., 2009). For example, the human conditions — livelihoods, infrastructure, institutions, social culture and behaviour, demography, geography and environment, as well as the disaster risk management system, mitigation, early warning, response and post-disaster recovery — all influence coastal vulnerability (Zou & Wei, 2010).

The health consequences and fatalities of climate change events are multifaceted — psychophysical, vector-water-food and airborne diseases and casualties (Portier et al., 2010). In some circumstances, the psychological and physical health impact of a disaster is worse for older people than for other groups (Haq et al., 2016; HelpAge International, 2015). Like the diverse health impacts, evidence shows that climate change also contributes to the increasing gap in livelihood and social inequalities among older people and their households in the affected community (CIEH, 2008; Friel, 2014; Paavola, 2017). Older people are more vulnerable to the negative health effects of climate change because of their physical decline or weakness. Studies have shown that extreme climate events have a greater impact on older people, leading to a higher incidence of flood-related morbidity and a higher mortality rate (Ahern et al., 2005). Flooding often results in higher fatality rates among older people than in other age groups, owing to direct causes such as drowning and secondary health effects such as hypothermia and cardiac difficulties (Leyva et al., 2017). For example, Kabir and Khan (2017) found that diarrhoea, skin diseases and mental health problems increased after cyclones. Age, gender, income and education level of household heads, as well as the number of children living in the family, had statistically significant effects on experiencing health problems before and after Cyclones SIDR and AILA. Extreme weather events such as Cyclones SIDR and AILA make life more difficult for coastal dwellers and increase the vulnerability of poor people, the elderly, children and women in society.

3.4 Social Protection and Vulnerability

The socio-economic resilience and adaptation capacities of older people may reduce their susceptibilities to extreme weather, which is dependent on their capabilities, capital assets and support systems. Socioeconomic policies and programmes such as contributory and non-contributory pensions, cash transfers, social food programmes, social protection, and other supports for income-generating activities increase resiliency and reduce the vulnerability of older people to extreme climate events (Zaidi, 2014). A growing number of extreme weather events are impacting the limited resources of older people in the affected country increasing the cost of disaster response, which has led researchers, international agencies, governments and local organisations to explore how social protection can address the consequences of extreme weather events in different countries. Social protection can increase household resilience by facilitating the anticipatory, absorbing, and adapting to covariate shocks of extreme weather events (Hebbar & Shebab, 2020).

Social protections help the poor and vulnerable to cope with crises and shocks, find jobs, improve productivity, invest in their children's health and education, and protect the ageing population. Social protection programmes empower people to be healthy, continue their education and seek opportunities to lift themselves and their families out of poverty (World Bank, 2022). Social protection measures such as safety nets, livelihood promotion, social funds to promote community-based adaptation, microfinance and microinsurance, skills development, supported migration and labour market interventions can help address the consequences of extreme weather events through prevention, protection and promotion strategies to reduce vulnerability (Aleksandrova, 2019; Kuriakose et al., 2013; World Bank, 2012).

The various conditions of older people that affect their ability to respond to extreme hazards include social protection, livelihood and resilience, self-protection, and socio-political and institutional networks, which can exacerbate or influence the extent of older people's vulnerability regardless of the type of hazard. Cannon (2008) included social protection in his vulnerability approach along with other components that capture all aspects of vulnerability to natural hazards: (i) livelihoods, their strength and resilience to hazards, which include income-generating activities and access to available resources; (ii) wellbeing and basic statuses such as physical and mental health, nutritional status, morale, sense of security and identity; (iii) capacity for self-protection; (iv) social protection; and (v) governance.

It is argued that these five elements are interrelated and the links between them are particularly important in understanding the determinants of vulnerability and designing policies to reduce them. The most important linkages are those that affect livelihood strength and social protection, both of which depend primarily on governance to determine how effective they are. The basic building block for determining a person's vulnerability is how adequate their livelihood is and how resilient they are to risk. The wellbeing and self-protection of a person are strongly determined by the strength of their livelihoods. The distribution of different livelihoods among certain groups of people depends heavily on the type of governance that influences the distribution of income and wealth that is needed to reduce vulnerability (Cannon, 2008). Conversely, the failure of the governance of formal and informal social protection can increase the vulnerability to extreme weather events in the long run. As Mallick et al. (2011) showed, 'dependence on relief and rehabilitation materials' makes older people more vulnerable to further disasters. In the long run, this increases poverty levels and puts them under pressure to remain in a 'vulnerability trap' once a disaster strikes.

Thus, older people are likely to be particularly vulnerable, especially where social protection is limited or non-existent, and they are at high risk from the impacts of climatic events. These programmes can ensure and protect their rights (to health and medical facilities, shelter and social security, food and nutrition, clothing, recreation, employment, etc.) as senior citizens in the community. Hence, social protection is increasingly seen as a means to mitigate vulnerability, build capacity, and meet the basic needs of older people in coastal regions for adaptation and recovery after extreme weather events.

3.5 Models and Approaches of Vulnerability

The concept and approach of vulnerability differ across disciplines; however, the fundamental meaning is exposure to threats with a limited adaptive capacity to handle these threats effectively (Siegel & Alwang, 1999). Though the concept of vulnerability was developed in the environmental sciences as an analytical concept related to the impact of natural and economic adversities on human populations, it is commonly used in health, social care, sociology, geography, economics, food security, and social work areas (Alwang et al., 2001; Brown et al., 2017; Wisner, 1993). However, each community of researchers has developed their own discrete conceptual designs and procedures using distinct languages (Crooks, 2009; Füssel, 2007). They have used various terms such as susceptibility, adaptive capacity, risk, hazards and threats to conceptualise vulnerability to extreme hazards (Adger, 2006; Burton et al., 2002). While climate scientists often consider vulnerability in terms of the effect of climate change and the impact of extreme weather events, social scientists tend to present it as a set of socioeconomic factors, which determine how people can deal with stress, risk, threats and change (Allen, 2003).

United Nations (2004) also identifies four different groups of vulnerability factors: physical, economic, social and environmental that increase the susceptibility of a community to the impact of extreme weather events. The physical factors denote the exposure of vulnerable elements of the area. Economic factors describe the financial resources of individuals, groups and communities; social factors such as educational attainment, security, access to basic human rights and good governance determine the well-being of individuals, populations and communities; and environmental factors provide information about the state of conditions in a region. Instead of external pressures, all these issues show characteristics of a weak system or community (Füssel, 2007).

Famine vulnerability and food security models have been developed, among others, to explain vulnerability to famine in the absence of food shortages or production losses (Sen, 1981). Vulnerability is defined in these models as a failure of entitlements and a lack of capabilities. The vulnerability model has also been used to identify and forecast different vulnerable groups and vulnerable regions based on the likelihood and consequences of hazards. The pressure and release model, based on the human ecology model, was developed to link discrete risks to the political economy of resources and normative disaster management and intervention (Pelling, 2003; Wisner et al., 2012). The physical and biological hazards create pressure and lead to vulnerability, while local geography and social differentiation create additional pressure from causes, leading to a cumulative increase in vulnerability.

The analysis of economic elements and social interactions in the sustainable livelihoods and poverty vulnerability paradigm is used to explain why people become or remain poor over time. However, Watts and Bohle (1993) conceptualised vulnerability in terms of risks: (1) the risk of being exposed to external shocks and hazards, (2) the risk of not being able to mobilise resources quickly enough to respond to hazards, and (3) the risk of experiencing the severe negative effects of shock, hazard and stress. Both environmental threats and the resources that people or groups mobilise to deal with these risks are described through this concept.

The hazard-of-place model offers a helpful insight into how vulnerability is considered under catastrophic conditions. In this model, various social elements such as socio-economic characteristics, risk perception, resilience and the built environment either mitigate or increase hazard potential (Cutter et al., 2003). Geographical features such as location, altitude and proximity also influence hazard potential. The combination of biophysical and social vulnerability thus gives the overall vulnerability of a place and serves to represent the overall situation by outlining the various factors that influence the vulnerability of a particular geographical region.

Human vulnerability to natural hazards (extreme weather events) is caused by a variety of social, economic, historical and political factors, all operating at different levels. However, the causes of vulnerability are primarily social and economic, rather than the result of differential exposure to climate and extreme hazards. Several factors, including access to resources, governance, culture and knowledge, that are responsible for local differences in vulnerability to extreme weather events and climate change. Access to resources is a critical factor influencing the ability of communities to plan for and respond to the impacts of climate change, and a lack of resources can double vulnerability. In addressing the impacts of climate change and extreme weather events, a lack of effective governance is both an acute and chronic challenge. Local or regional cultural and historical circumstances may limit possible responses to climate change threats (Thomas et al., 2018).

Laila (2013) assessed the social vulnerability of the coastal societies of Bangladesh to understand their social situations based on the limited natural resources through the IPCC components — exposure, sensitivity and adaptive capacity. The geographical location and housing conditions of the coastal communities studied were more vulnerable, with the elderly and disabled among the most vulnerable groups, as indicated by the exposure component. While sanitation, water quality, livelihood, income and food security are used to assess sensitivity, telecommunication tools such as radio and television are considered to reflect adaptive capacity. The coastal communities of Bangladesh are highly dependent on natural resources such as water, have few livelihood options and are poorly prepared for the negative impacts of climate change.

3.6 The Integrative Framework for Vulnerability

The approaches of vulnerability to climate change and extreme weather events can be broadly divided into biophysical, social and integrated perspectives. In the biophysical approach, vulnerability is an adverse biophysical state that considers the susceptibility and deterioration of the environment that affects the potential consequences for people's lives. The biophysical approach focuses on the distribution of hazardous situations, settlement in unsafe regions, and the extent of losses associated with a particular extreme event. While this view can provide a comprehensive understanding of the physical processes that lead to vulnerability, it is limited because it ignores the socio-economic, cultural and political factors that must be considered when assessing vulnerability (Cardona, 2004). In contrast, vulnerability is viewed from a social perspective as a socially constructed condition that has its origins in particular socio-political structures, historical and economic progress and can affect people in the community, placing them in vulnerable situations (Brooks, 2003; Cutter, 1996). Vulnerability in this perspective is seen as the intrinsic condition of people, communities, social groups or the whole social system. The integrated approach to vulnerability combines the two views by integrating the biophysical and social dimensions of vulnerability.

The vulnerability of older people to extreme weather events thus can be determined by the adverse effects of exposure, sensitivity and capacity, which depend on their geographical location, socio-economic conditions and access to assets and resources, social networks and socio-state support system. An integrated framework is, therefore, required to explore their vulnerability, incorporating both physical and social dimensions. In this book, the vulnerability of older people to extreme weather events and the role of social protection in Bangladesh are explored using an integrated conceptual framework, the interrelated factors of which are shown in the diagram below (Table 3.1):

3.6.1 The Concepts in the Framework

The concept of *exposure* in the framework defines the nature, magnitude and rate of change and variation in climate to which a system is exposed. Typically, exposure includes factors such as temperature, precipitation, climatic water balance as well as extreme weather events. Exposure refers to the exposure of a system of interest to stimuli acting on it. This can be conceptualised as climate variability and/or the various changes in the climate system that often cause concern to stakeholders: Temperature increase, precipitation variability and change (including extremes), or changes in the frequency or intensity of tropical cyclones (IPCC, 2001). Various studies claim that people living in the coastal region are more exposed to extreme weather events. Thus, extreme weather events such as cyclones and floods were considered as exposure in this study.

Table 3.1 Domain and components of the vulnerability of older people

Domain in the framework	Components analysed in this study
Exposure	Extreme weather events: • Cyclones • Floods due to storm surges
Sensitivity	• Geophysical location of settlements • Housing conditions • Conditions of the cyclone shelter
(Potential) impacts	• Livelihood insecurity • Loss of resources • Damage to dwellings and living • Food insecurity • Deterioration of health
Adaptive capacity	• Socioeconomic status • Employment status • Living facilities
Social protection	• Formal social protection • Informal social protection

Source Developed by authors (integrating Cutter et al., 2003; IPCC, 2001, 2014; Turner II et al., 2003 models).

An ***extreme weather event*** can be considered a state of weather that is outside of normal weather patterns. The IPCC report by Seneviratne et al. (2012) defines an 'extreme climate or weather event' or 'climate extreme' as "the occurrence of a value of a weather or climate variable above (or below) a threshold value near the upper (or lower) ends of the range of observed values of the variable." These weather events include heavy rainfall and flooding, cyclones, droughts, heat waves, extreme cold and forest fires. Each of these events can potentially impact the socioeconomic conditions of society and the integrity of supply and distribution systems, which may increase the vulnerability of any affected community (Khan et al., 2015). Extreme weather events can affect populations and the environment in different ways depending on their sensitivity to extreme situations and their resilience to recover from stress. The cyclones and floods that occurred before 2019 in coastal Bangladesh are considered in this study to explore their impacts on the vulnerability of older people.

A ***cyclone*** refers to a storm or system of winds that rotates about a centre of low atmospheric pressure, advances at a speed of 20–30 miles (about 30–50 km) an hour, and often brings heavy rain (Merriam-Webster, 2022). A tropical cyclone, also called a typhoon or hurricane, is an intense circular storm formed over warm tropical oceans and characterised by low atmospheric pressure, strong winds and heavy rain. It produces winds that exceed 119 km (74 miles) per hour. In extreme cases, winds can exceed 150 miles (240 km) per hour, and gusts can reach 200 miles (320 km) per hour. Accompanying these strong winds are torrential rains and a devastating phenomenon known as storm surge. This is a rise in the sea surface that can be up to 20 ft (6 m) above normal. Such a combination of strong winds and water makes cyclones a serious threat to coastal areas in tropical and subtropical regions of the world.

A ***flood*** is an overflow of water that inundates normally dry land. In the sense of "flowing water", the word can also be applied to the inflow of the tides. The inundation of a normally dry area by rising water in an existing waterway, such as a river, stream or drainage ditch. The accumulation of water at or near the site where the rain fell. A flood is a longer-term event than a flash flood: it can last for days or weeks. Again, a flash flood is an overflow of water caused by heavy or excessive rainfall within a short period, usually less than 5 h. Flash floods are usually characterised by raging torrents after heavy rains that roll through riverbeds, urban streets or mountain gorges, taking everything with them. They can occur within minutes or a few hours after excessive rainfall.

Sensitivity determines the degree to which a system is adversely (or beneficially) affected by a given climate (change) exposure. Sensitivity may be determined by (a) natural/physical factors of a system, or related to human land management activities and infrastructure and societal factors. Sensitivity refers to the responsiveness of a system to extreme climate hazards. This is often conceptualised as a dose-response model — the more sensitive a system is, the greater the rate or magnitude of an adverse response to a particular hazard. Sensitivity can vary greatly from system to system, sector to sector or population to population; for example, how much land could be flooded by sea level rise or how much crop yields could change (IPCC, 2001). The human–environment conditions of the system determine its sensitivity to any set of exposures. These conditions include both social and biophysical capital that influence the existing coping mechanisms, which take effect as the impacts of the exposure are experienced, as well as those coping mechanisms adjusted or created because of the experience (Turner II et al., 2003). This study considered the geophysical location of settlements, housing conditions and conditions of cyclone shelters as sensitivity factors for older people in the regions.

The potential ***impacts*** of extreme weather events are defined as the positive, harmful or mitigable effects that may disrupt the normal functioning of the system and thus constitute a 'climate threat'. The potential impacts are measured as a function of the exposure and sensitivity of the system. It is the combined effect of exposure and sensitivity without additional adaptation measures, such as riverbank erosion, and loss of income due to torrential rains. The impacts depend on access to resources that could help in responding to threats and exposures. This study assessed livelihood insecurity, loss of resources and damage to dwellings and housing, food insecurity, deterioration of health, disruption of relationships and mutual assistance as impacts of extreme weather events — cyclones and floods.

Adaptive capacity denotes the ability of a system to change so that it is better able to cope with its exposure and/or sensitivity to climatic influences (IPCC, 2001). Adaptive capacity refers to the ability of a human or natural system to adjust to extreme weather events, including climate variability and extremes; prevent or moderate potential damages; take advantage of opportunities; or cope with the consequences (Kuriakose et al., 2013). The adaptive capacity of a household, region, or country depends on its stock of financial and economic resources, access to technology, information and skills, infrastructure assets, institutional assets, and degree of equity. Adaptive capacity describes the ability of a society, social group or individuals to actively adapt to climate change, climate variability and climate extremes by mitigating potential damages, seizing opportunities or coping with the consequences. Various studies, theories and models of vulnerability include adaptive capacity as a component that refers to the ability or capacity of a system, person or community to adapt to the impacts of extreme weather events. Adaptive capacity has a significant impact on the vulnerability of communities and regions exposed to the impacts and hazards of extreme weather events. This study considered socioeconomic status, employment status, social networks and housing and living facilities as the capacity of older people to address their vulnerability to extreme weather events.

Social protection policies and programmes are now considered effective tools to protect vulnerable groups from livelihood risks and hazards (World Bank, 2001). Social protection can be an important strategy by cushioning the impact of shocks and stress through income replacement and building resilience by reducing exposure to extreme weather events, providing incentives/assistance to retrofit homes to withstand tropical cyclones, or implementing public works programmes to restore income to cope with the impact of extreme events. The current study looks at formal and informal social protection programmes and assesses how they address older people's vulnerability to extreme weather events. Social protection programmes were first integrated into the vulnerability framework to assess how they mitigate the vulnerability of older people by reducing their sensitivity and/or increasing their capacity of older people (as shown in Fig. 3.1). During extreme weather events in the study region, a variety of social protection programmes helped older people and their families, including the provision of food, clothing, in-kind transfers, housing, household amenities, cash transfers, and employment guarantee schemes in response to the disaster. The current study considers the contribution of social protection to address the vulnerability of older people to extreme weather events. Thus, social protection, in this study, has been considered to include all formal and informal initiatives taken by the government and non-government sources to build capacity and access to meet the basic needs of older people to enhance socio-economic conditions and mitigate vulnerability to extreme weather events (Table 3.2).

3.7 Conclusion

There is consensus that extreme weather events induced by climate change have diverse impacts on humans and the environment. This takes place directly through changing temperatures, precipitation and the occurrence of heat waves, floods, droughts, fires and, indirectly, by ecological disruptions that bring crop failures and

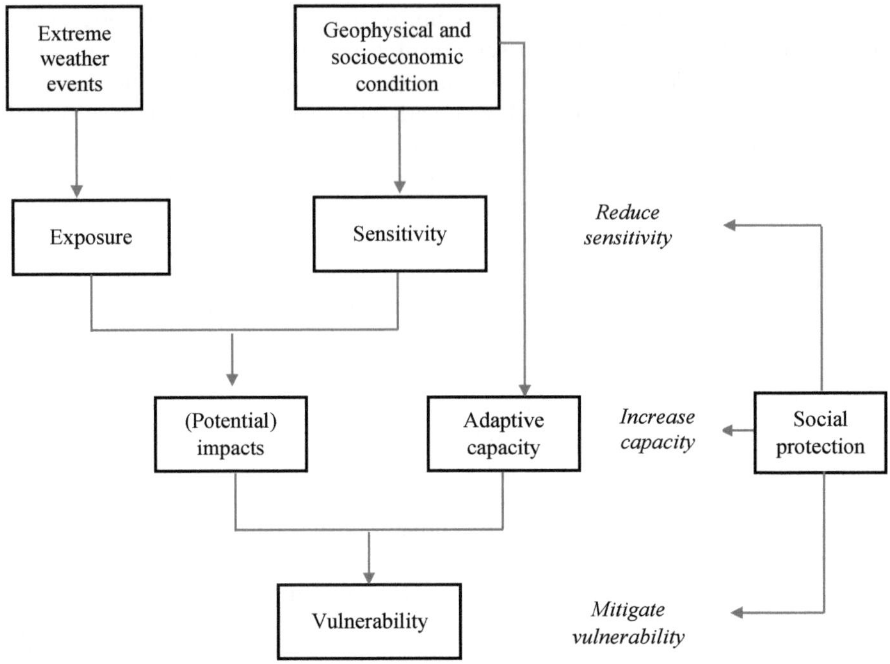

Fig. 3.1 Integrated framework for vulnerability and social protection. *Source* Developed by authors (from Cutter et al., 2003; IPCC, 2001, 2014; Turner II et al., 2003; World Bank, 2001)

Table 3.2 The social protection programmes considered in this study

Categories of Social Protection	Role in Capacity Building	Contribution to access needs
Formal Social Protection: • Old Age Allowance • Widow Allowance • Public Works • Relief Programmes **Informal Social Protection:** • Assistance from • Volunteer groups and NGOs • Neighbours and friends • Family members and relatives	• Anticipatory capacity • Protective capacity • Preventive capacity • Adaptive capacity • Absorptive and recovery capacity • Rehabilitative capacity	• Food • Self-employment • Medicine • Clothing • Housing • Social Networks

Source Adapted from World Bank (2001, 2012), Kuriakose et al. (2013), Cannon (2008) and Department of Social Services, Government of the People's Republic of Bangladesh. https://dss.gov.bd/

shift the patterns of disease vectors, or social responses like displacement of populations. Bangladesh is one of the disaster-prone areas of the world, which is increasingly endangered through climate change that affects the socio-economic conditions and health of different social groups, especially of the older population. Older people are at greater risk from storms, floods, heat waves, and other extreme events. Besides the extreme events and socio-economic vulnerabilities, climate change is also responsible for the spread of various diseases among older people.

Various extreme weather events have direct and indirect impacts on the vulnerability of older people in Bangladesh as well as in other countries. Some studies have examined the vulnerability of people and places to extreme weather events in different ways. However, despite the extensive research on natural disasters and extreme weather events, very few studies focus on older people and their vulnerability. The studies on extreme weather events, climate change and disasters in Bangladesh hardly address the vulnerability of older people and the role of social protection in a comprehensive perspective. Using an integrated framework, this book investigates the vulnerability of older people to extreme weather events — cyclones and floods — in the coastal region of Bangladesh, encompassing both biophysical and socio-economic aspects of vulnerability and social protection.

References

Adger, W. N. (2006). Vulnerability. *Global Environmental Change, 16*(3), 268–281. https://doi.org/10.1016/j.gloenvcha.2006.02.006

Ahern, M., Kovats, R. S., Wilkinson, P., Few, R., & Matthies, F. (2005). Global health impacts of floods: Epidemiologic evidence. *Epidemiologic Reviews, 27*(1), 36–46.

Aleksandrova, M. (2019). Principles and considerations for mainstreaming climate change risk into national social protection frameworks in developing countries. *Climate and Development, 12*(1), 1–10.

Allen, K. (2003). Vulnerability reduction and the community-based approach. In M. Pelling (Ed.), *Natural disasters and development in a globalising world* (pp. 170–184). Routledge.

Alwang, J., Siegel, P. B., & Jørgensen, S. L. (2001). *Vulnerability: A view from different disciplines* (Social Protection Discussion Paper No. 0115). World Bank.

Ball, N. (1975). The myth of the natural disaster. *The Ecologist, 5*(10), 368–369.

Bates, B. C., Kundzewicz, Z. W., Wu, S., & Palutikof, J. P. (Eds.). (2012). *Climate change and water*. IPCC Secretariat.

Blaikie, P., Cannon, T., Davis, I., & Wisner, B. (2014). *At risk. Natural hazards, people's vulnerability and disasters*. Routledge.

Brooks, N. (2003). *Vulnerability, risk and adaptation: A conceptual framework* (Tyndall Centre Working Paper No. 38). Tyndall Centre for Climate Change Research.

Brooks, N., Brown, K., & Grist, N. (2009). Development futures in the context of climate change: Challenging the present and learning from the past. *Development Policy Review, 27*(6), 741–765. https://doi.org/10.1111/j.1467-7679.2009.00468.x

Brown, K., Ecclestone, K., & Emmel, N. (2017). The many faces of vulnerability. *Social Policy and Society, 16*(3), 497–510.

Burton, I., Kates, R. W., & White, G. F. (1978). *The environment as hazard*. Oxford University Press.

Burton, I., Kates, R. W., & White, G. F. (1993). *The environment as hazard* (2nd ed.). Guilford Publications.
Burton, I., Huq, S., Lim, B., Pilifosova, O., & Schipper, E. L. (2002). From impacts assessment to adaptation priorities: The shaping of adaptation policy. *Climate Policy, 2*(2–3), 145–159. https://doi.org/10.3763/cpol.2002.0217
Calzadilla, A., Pauli, F., & Roson, R. (2006). *Climate change and extreme events: An assessment of economic* (No. 825-2016-55066). Fondazione Eni Enrico Mattei (FEEM).
Cannon, T. (2008). *Reducing people's vulnerability to natural hazards communities and resilience* (Research Paper No. 2008.34). UNU-WIDER.
Cardona, O. D. (2004). The need for rethinking the concepts of vulnerability and risk from a holistic perspective: A necessary review and criticism for effective risk management. In G. Frerks & D. Hilhorst (Eds.), *Mapping vulnerability: Disasters, development and people* (pp. 37–51). Earthscan.
Cardona, O. D. (2006). A system of indicators for disaster risk management in the Americas. In J. Birkmann (Ed.), *Measuring vulnerability to natural hazards—Towards disaster resilient societies* (pp. 181–197). United Nations University Press.
Chambers, R. (1989). Editorial introduction: Vulnerability, coping and policy. *IDS Bulletin, 20*(2), 1–7.
Chartered Institute of Environmental Health. (2008). *Older people and climate change: A review.* Chartered Institute of Environmental Health.
Clark, G. E., Moser, S. C., Ratick, S. J., Dow, K., Meyer, W. B., Emani, S., Jin, W., Kasperson, J. X., Kasperson, R. E., & Schwarz, H. E. (1998). Assessing the vulnerability of coastal communities to extreme storms: The case of Revere, MA., USA. *Mitigation and Adaptation Strategies for Global Change, 3*(1), 59–82.
Crooks, D. (2009). *Development and testing of the elderly social vulnerability index (ESVI): A composite indicator to measure social vulnerability in the Jamaican elderly population* (PhD dissertation). Florida International University, Miami, FL, USA.
Curtis, S., Fair, A., Wistow, J., Val, D. V., & Oven, K. (2017). Impact of extreme weather events and climate change for health and social care systems. *Environmental Health, 16*(1), 23–32.
Cutter, S. L. (1996). Vulnerability to environmental hazards. *Progress in Human Geography, 20*(4), 529–539.
Cutter, S. L., Boruff, B. J., & Shirley, W. L. (2003). Social vulnerability to environmental hazards. *Social Science Quarterly, 84*(2), 242–261.
Dwyer, A., Zoppou, C., Nielsen, O., Day, S., & Roberts, S. (2004). Quantifying social vulnerability: A methodology for identifying those at risk to natural hazards. (Geoscience Australia Record 2004/14). Canberra: Geoscience Australia. https://www.ga.gov.au/bigobj/GA4267.pdf
Emmerson, C., & Muriel, A. (2008). *Poverty in old age: An update.* Institute for Fiscal Studies.
Feist, H. (2016, March/April). Growing old in a changing climate. *Australian Ageing Agenda*, p. 12.
Filiberto, D., Wethington, E., Pillemer, K., Wells, N., Wysocki, M., & Parise, J. T. (2009). Older people and climate change: Vulnerability and health effects. *Generations, 33*(4), 19–25. Retrieved December 3, 2020, from https://www.researchgate.net/publication/265265398_Older_People_and_Climate_Change_Vulnerability_and_Health_Effects
Fineman, M. A. (2008). The vulnerable subject: Anchoring equality in the human condition. *Yale Journal of Law and Feminism, 20*, 1.
Fineman, M. A. (2012). 'Elderly' as vulnerable: Rethinking the nature of individual and societal responsibility. *Elder Law Journal, 20*(1), 71–112. https://ssrn.com/abstract=2088159
Fineman, M. A. (2016). Equality, autonomy, and the vulnerable subject in law and politics. In M. A. Fineman & A. Grear (Eds.), *Vulnerability* (pp. 25–40). Routledge.
Forbes-Mewett, H., & Nguyen-Trung, K. (2019). Defining vulnerability. In H. Forbes-Mewett (Ed.), *Vulnerability in a mobile world* (pp. 5–27). Emerald Publishing Limited.
Ford, J. D., Pearce, T., McDowell, G., Berrang-Ford, L., Sayles, J. S., & Belfer, E. (2018). Vulnerability and its discontents: The past, present, and future of climate change vulnerability research. *Climatic Change, 151*(2), 189–203.

Friel, S., Berry, H., Dinh, H., O'Brien, L., & Walls, H. L. (2014). The impact of drought on the association between food security and mental health in a nationally representative Australian sample. *BMC Public Health, 14*, 1102. https://doi.org/10.1186/1471-2458-14-1102 (Accessed November 30, 2020)

Fritzsche, E., et al. (2014). The social dimension of vulnerability to environmental change. *International Journal of Disaster Risk Reduction, 9*, 198–207.

Füssel, H.-M. (2007). Vulnerability: A generally applicable conceptual framework for climate change research. *Global Environmental Change, 17*(2), 155–167.

Grundy, E. (2006). Ageing and vulnerable older people: European perspectives. *Ageing and Society, 26*(1), 105–134.

Haq, G., & Gutman, G. (2020). Climate resilience and older people. In D. Gu & M. Dupre (Eds.), *Encyclopedia of gerontology and population aging* (pp. 1–4). Springer.

Haq, G., Whitelegg, J., & Kohler, M. (2008). Growing old in a changing climate: Meeting the challenges of an ageing population and climate change. Stockholm Environment Institute. Retrieved from https://www.sei.org/mediamanager/documents/Publications/Future/climate_change_growing_old.pdf

Hebbar, M., & Shebab, S. (2020). *Shock-responsive social protection in Bangladesh*. OPM and UK aid.

Heitzmann, K., Canagarajah, R. S., & Siegel, P. B. (2002). *Guidelines for assessing the sources of risk and vulnerability* (WB Social Protection Discussion Paper No. 0218). World Bank.

HelpAge International. (2015). *Climate change in an ageing world*. HelpAge International.

Intergovernmental Panel on Climate Change. (2001). *Climate change 2001: The scientific basis*. Contribution of Working Group I to the Third Assessment Report (881 pp.). Cambridge University Press. Retrieved from https://www.ipcc.ch/report/ar3/wg1/

Intergovernmental Panel on Climate Change. (2014). *Climate change 2014: Impacts, adaptation, and vulnerability. Part A: Global and sectoral aspects*. Cambridge University Press.

Kabir, R., & Khan, H. T. A. (2017). Study on the health status of coastal people in Bangladesh after cyclones Sidr and Aila. *European Scientific Journal, 13*(15), 329–342.

Kelman, I., Gaillard, J. C., Lewis, J., & Mercer, J. (2016). Learning from the history of disaster vulnerability and resilience research and practice for climate change. *Natural Hazards, 82*(1), 129–143.

Kelly, P. M., & Adger, W. N. (2000). Theory and practice in assessing vulnerability to climate change and adaptations. *Climatic Change, 47*(4), 325–352.

Khan, S. J., Deere, D., Leusch, F. D., Humpage, A., Jenkins, M., & Cunliffe, D. (2015). Extreme weather events: Should drinking water quality management systems adapt to changing risk profiles? *Water Research, 85*, 124–136.

Kreager, P. (2006). Migration, social structure and old-age support networks: A comparison of three Indonesian communities. *Ageing and Society, 26*(1), 37–60.

Kuriakose, A. T., Heltberg, R., Wiseman, W., Costella, C., Cipryk, R., & Cornelius, S. (2013). Climate-responsive social protection. *Development Policy Review, 31*(o19), 19–34.

Laila, F. (2013). *Assessment on social vulnerabilities to climate change—A study on the south-western coastal region of Bangladesh* (Master's thesis No. 154). Uppsala University, Uppsala, Sweden.

Levy, N., Clark, S., & O'Neill, E. (2016). *Living longer, working longer* (IZA DP No. 10235). IZA Institute of Labor Economics.

Leyva, E. W. A., Beaman, A., & Davidson, P. M. (2017). Health impact of climate change in older people: An integrative review and implications for nursing. *Journal of Nursing Scholarship, 49*(6), 670–678.

Liverman, D. M. (1990). Vulnerability to global environmental change. In R. E. Kasperson, K. Dow, D. Golding, et al. (Eds.), *Understanding global environmental change: The human dimensions* (pp. 27–43). National Academy Press.

Malak, M. A., Sajib, A. M., Quader, M. A., & Anjum, H. (2020). "We are feeling older than our age": Vulnerability and adaptive strategies of aging people to cyclones in coastal Bangladesh. *International Journal of Disaster Risk Reduction, 48*, Article 101595.
Mallick, B., Rahaman, K. R., & Vogt, J. (2011). Social vulnerability analysis for sustainable disaster mitigation planning in coastal Bangladesh. *Disaster Prevention and Management, 20*(3), 220–237.
McLaughlin, D. K., & Jensen, L. (2000). The rural elderly: A demographic portrait. *Annals of the American Academy of Political and Social Science, 5*(165), 87–111.
Merriam-Webster. (2022). *Cyclone*. Retrieved April 16, 2022, from https://www.merriam-webster.com/dictionary/cyclone
Obi, J. N., Enete, A. A., & Munonye, J. O. (2021). Farm households' levels of vulnerability to extreme weather events in South-Eastern Nigeria. *African Journal of Food, Agriculture, Nutrition and Development, 21*(2), 17450–17463.
O'Keefe, P., Westgate, K., & Wisner, B. (1976). Taking the naturalness out of natural disaster. *Nature, 260*, 566–567. https://doi.org/10.1038/260566a0
Otto, I. M., Reckien, D., Reyer, C. P., Marcus, R., Le Masson, V., Jones, L., & Serdeczny, O. (2017). Social vulnerability to climate change: A review of concepts and evidence. *Regional Environmental Change, 17*(6), 1651–1662.
Oven, K., Wistow, J., & Curtis, S. (2019). Older people and climate change: Vulnerability and resilience to extreme weather in England. In L. Reyes Mason & J. Rigg (Eds.), *People and climate change: Vulnerability, adaptation, and social justice* (pp. 68–83). Oxford University Press.
Paavola, J. (2017). Health impacts of climate change and health and social inequalities in the UK. *Environmental Health, 16*, 61–68.
Pelling, M. (2003). *The vulnerability of cities: Natural disasters and social resilience*. Earthscan.
Portier, C. J., Thigpen Tart, K., Carter, S. R., Dilworth, C. H., Grambsch, A. E., Hess, J., Gohlke, J., Hess, J., Howard, S. N., Luber, G., Lutz, J. T., Maslak, T., Prudent, N., Radtke, M., Rosenthal, J. P., Rowles, T., Sandifer, P. A., Scheraga, J., Schramm, P. J., Strickman, D., et al. (2010). *A human health perspective on climate change: A report outlining the research needs on the human health effects of climate change*. Environmental Health Perspectives and the National Institute of Environmental Health Sciences.
Rank, M. R. (2008). *Asset building over the life course: Final report* US Department of Health and Human Services, Office of the Assistant Secretary for Planning and Evaluation (DHHS/ASPE).
Raymond, C., Horton, R. M., Zscheischler, J., Martius, O., AghaKouchak, A., Balch, J., & White, K. (2020). Understanding and managing connected extreme events. *Nature Climate Change, 10*(7), 611–621.
Schröder-Butterfill, E., & Marianti, R. (2006). A framework for understanding old-age vulnerabilities. *Ageing and Society, 26*(1), 9–35.
Sen, A. (1980). *Levels of inequality: Notes on the measurement of poverty*. Institute of Development Studies.
Sen, A. (1981). *Poverty and famines: An essay on entitlement and deprivation*. Oxford University Press.
Seneviratne, S. I., Field, C. B., Barros, V., Stocker, T. F., Qin, D., Alexander, L. V., & Zwiers, F. (2012). Changes in climate extremes and their impacts on the natural physical environment. In C. B. Field et al. (Eds.), *Managing the risks of extreme events and disasters to advance climate change adaptation (SREX)* (pp. 109–230). Cambridge University Press.
Siegel, P. B., & Alwang, J. (1999). An asset-based approach to social risk management: A conceptual framework (Social Protection Discussion Paper No. 9926). The World Bank. https://documents.worldbank.org/en/publication/documentsreports/documentdetail/837121468765606678/an-asset-based-approach-to-social-risk-management-a-conceptual-framework
Thomas, K., Hardy, R. D., Lazrus, H., Mendez, M., Orlove, B., Rivera-Collazo, I., Roberts, J. T., Rockman, M., Warner, B. P., & Winthrop, R. (2018). Explaining differential vulnerability to climate change: A social science review. *WIREs Climate Change, 10*(2), e565.

Tierney, K., Lindell, M., & Perry, R. (2001). *Facing the unexpected: Disaster preparedness and response in the United States*. Joseph Henry Press.

Timmerman, P. (1981). *Vulnerability, resilience and the collapse of society. A review of models and possible climatic applications*. Institute for Environmental Studies, University of Toronto.

Turner, B. L., II, Kasperson, R. E., Matson, P. A., McCarthy, J. J., Corell, R. W., Christensen, L., Eckley, N., Kasperson, J. X., Luers, A., Martello, M. L., Polsky, C., Pulsipher, A., & Schiller, A. (2003). A framework for vulnerability analysis in sustainability science. *Proceedings of the National Academy of Sciences, 100*(14), 8074–8079.

United Nations. (2003). *Report on the world social situation 2003: Social vulnerability—sources and challenges* pp. 1–95. Department of Economic and Social Affairs, Division for Social Policy and Development. Retrieved December 20, 2019, from https://www.un.org/esa/socdev/rwss/docs/2003/fullreport.pdf

United Nations. (2004). *Living with risk: A global review of disaster reduction initiatives*. United Nations.

United Nations Development Programme. (2004). *Reducing disaster risk: A challenge for development*. United Nations.

UNISDR. (2015). *Sendai framework for disaster risk reduction 2015–2030*. United Nations.

Van Eeuwijk, P. (2006). Old-age vulnerability, ill-health and care support in urban areas of Indonesia. *Ageing and Society, 26*(1), 61–80.

Van Westen, C. (2014). *Conceptual and methodological framework for the analysis of vulnerability to climate change*. U.S. Agency for International Development.

Virokannas, E., Liuski, S., & Kuronen, M. (2020). The contested concept of vulnerability—A literature review: Vulnerability-käsitteen kiistanalaiset merkitykset-systemaattinen kirjallisuuskatsaus. *European Journal of Social Work, 23*(2), 327–339.

Watts, M. J., & Bohle, H.-G. (1993). The space of vulnerability: The causal structure of hunger and famine. *Progress in Human Geography, 17*(1), 43–67.

Wisner, B. (1993). Disaster vulnerability: Scale, power and daily life. *GeoJournal, 30*(2), 127–140.

Wisner, B. (2016). Vulnerability as concept, model, metric, and tool. In *Oxford research encyclopaedia of natural Hazard science (pp. 1–51)*.. Oxford University Press. https://doi.org/10.1093/acrefore/9780199389407.013.25

Wisner, B., & Luce, H. R. (1993). Disaster vulnerability: Scale, power and daily life. *GeoJournal, 30*(2), 127–140.

Wisner, B., Blaikie, P., Cannon, T., & Davis, I. (2012). *At risk: Natural hazards, people's vulnerability and disasters*. Routledge.

World Bank. (2001). *Social protection sector strategy: From safety nets to social policy*. World Bank.

World Bank. (2012). *Social protection for a changing Asia*. World Bank Publications.

World Bank. (2022). *The World Bank in Social Protection*. Retrieved April 2, 2022, from https://www.worldbank.org/en/topic/socialprotection/overview#1

Zaidi, A. (2014). *Life cycle transitions and vulnerabilities in old age: A review* (Occasional Paper). UNDP Human Development Report Office.

Zou, L., & Wei, Y.-M. (2010). Driving factors for social vulnerability to coastal hazards in Southeast Asia: Results from the meta-analysis. *Natural Hazards, 54*(3), 901–929.

[illegible]

[illegible]

[illegible]

Chapter 4
Geographic and Socio-economic Vulnerabilities of Older People

4.1 Introduction

This chapter analyses the geophysical and Socioeconomic vulnerabilities of older people in the coastal region of Bangladesh. These vulnerabilities may be due to the geophysical location of the residential settlement, or they may be socio-economic in nature, such as lack of employment and income, poverty, poor housing conditions and dependence on others. The geological location and settlement pattern of the older people in the region make them more vulnerable to extreme weather events than those living on the higher plains. In the past, there have been numerous cyclones and tropical storms that have directly threatened the study area. The chapter interprets and discusses various external and internal factors such as settlement location and housing structure, poverty, employment, available resources, dependency, food insecurity and deteriorating health status. The analysis shows that older people in coastal Bangladesh have been severely affected by extreme weather events, and they believe that the same or more brutal conditions are highly likely to recur due to their poor socio-economic condition, location of their housing area and lack of a proper support system. Every extreme event, especially cyclones and floods due to storm surges, destroyed or significantly damaged their livelihoods and negatively affected their lives, making them vulnerable.

4.2 Geophysical Location and Settlements

Geophysical vulnerability considers geographical location and settlement patterns, as well as physical structures, including infrastructure in hazardous areas or with insufficient resilience or vulnerability to the impacts of extreme weather events (Cardona et al., 2012; UNISDR, 2004). The geographical locations where older people and their families live, as well as their housing structures, are highly vulnerable to various

M. I. Ali et al., *Extreme Weather Events and Vulnerability of Older People*,
https://doi.org/10.1007/978-3-032-11256-9_4

extreme weather events, especially cyclones and floods. In addition, older people are more vulnerable to poor geophysical and site-specific infrastructures due to their limited mobility, poor health, and limited access to key resources such as transport, medicine, water and food. Extreme weather events are predictable, and their impacts are entrenched in the coastal region of Bangladesh, pushing older people and their households into socio-economic vulnerability (Alauddin & Rahman, 2013).

The vulnerability of older people in the area is, to some extent, due to the uneven geographical environment and location of settlements. The study area is surrounded by deep water bodies on three sides: the Bishkhali, Haringhata and Baleshwar rivers are directly connected to the sea – the Bay of Bengal. Since most of the houses are close to the rivers and on the embankment, they may find it difficult to walk during extreme weather events due to uneven surfaces. Most of the older people said that their ancestors established the settlement in the area about 150 to 200 years ago after getting some land from the government of the time. As an older farmer said:

> [My] grandfather's family was in the Mathbaria area [an Upazila of Pirojpur district in Bangladesh]. My father and uncle came here [Patharghata - the study area] with him. My grandfather had a small amount of land, but he had six sons. We lived here after getting some land from the government. We [our family] came here more than 150 years ago (75-year-old small farmer, Ismail, Padma village).

It seems that 200 years ago, there were no settlements in the area, and the ancestors of the current residents probably founded this vulnerable settlement. Their grandparents came to this area from different parts of the country and from nearby districts to escape poverty and hunger. According to some locals, their grandparents had enough land properties in their homeland before coming here, but due to poverty and natural events such as floods, it was difficult for them to properly cultivate these landholdings and pay taxes on them. Thus, they fell into extreme poverty, which forced them to sell their land to survive and move to this coastal region. An 80-year-old Rashidul from Haritana village narrated that:

> My parents lived in the Mathbaria Upazila of the Pirojpur district. His parents were farmers. They had enough farmland [from their parents]. My grandparents once owned a lot of cultivable land, which had to be sold at an auction because they had not paid the land tax. Later, they moved from there to this area.

These statements were also corroborated by others. Most of the people living in the study area were from the Mathbaria and Pirojpur regions. The government gave them a piece of land that made them settle in this disaster-prone area. A 62-year-old man, Alamgir Hossain of Badurtala village, explained:

> My father came to Patharghata because he was allotted land here [by the government]. He came here because there was a severe [economic] crisis in Mathbaria at that time; rice was hardly grown there.

They had difficulty growing crops because the Patharghata region was part of the Sundarbans mangroves. The area was covered with forests and jungles as it was adjacent to the Sundarbans on the Bay of Bengal. People who came to this region first had to clear the jungle to open up a suitable area for living and cultivation. The land here was comparatively cheaper than flat land in nearby districts as the area was covered by deep jungles with wetlands. For instance, Ahammed, an 80-year-old widower, said his grandfather who had his ancestral home in the north in Mathbaria near the Jhalakathi district, settled in this area because he managed to purchase a piece of land here at a cheaper price.

Similarly, Debnath, a casual carpenter from Haritana village, and his father were born here, his grandfather was from another district, Jhalakathi, in the country.

Debnath commented that places that are more easily damaged are particularly vulnerable to extreme weather events such as sea level rise, storm surges, floods, and cyclones. Older people are particularly vulnerable to changes in the frequency or severity of cyclones and floods. Most houses are located near rivers that flow into the sea. The riverbanks are weak and can be destroyed at the first hit of a cyclone. As most of the elderly, their families and neighbours live on the coast, they are vulnerable to cyclones and floods. Enam, an 80-year-old widower, lives in a house that is on the bank of the river, only half a kilometre away from the sea. Some elderly people had houses that were located outside the embankment and very close to the rivers that are directly connected to the Bay of Bengal. Their houses are also highly exposed to the initial impact of the cyclone, making them even more vulnerable to extreme weather events. Asadul, a small farmer and casual labourer from Ruhita village, said:

> From my house, it is not far from the Ganga [river]. It takes me two to three minutes to reach the Ganga. My house is right next to the river. Our area is exposed to flooding even when there is a slight breeze. Whenever it is cloudy, I panic, and the fear and anxiety of disaster come to my mind. I was one of the sufferers of the Cyclone SIDR in 2007.

The entire study area and surrounding regions are prone to extreme weather events due to their location. The villages of Charlathimara and Padma are located on the banks of the Haringhata River, one of the major coastal rivers of the Ganges-Padma system, which flows into the Bay of Bengal. Ruhita village is on the boundary between the Haringhata and Bishkhali rivers. The dam near Charlathimara, Padma and Ruhita broke in 2007 due to the strong flood waves of cyclone SIDR and has not been properly rebuilt to date. Most elderly people in the area said that the dam was not strong enough to protect against the onslaught of water during cyclones and tidal waves. Some places were likely to face challenges as their physical infrastructure, including roads and embankments, was inadequate to protect against floods and the settlements were vulnerable to future extreme weather events (Kabir, 2009). Some older residents said that the dam was built with substandard quality materials and could break again if it is hit by cyclones. If this happens, flooding may occur, threatening the socio-economic livelihood of the elderly in the affected area. An older resident, Amena, said:

> They [the construction authority] built the roads [embankment]. But would it exist? No, it would not. When the water comes, everything will be damaged. [...] When the month of Baishakh [the first month of the Bangla New Year and the season of the cyclone, Kalbaishakhi] comes, I am scared and cannot sleep. In the name of Allah, I wish the wind would come and drive everything away.

As participants made clear in their responses, the underlying geophysical characteristics and location of the region near a large body of water increase the vulnerability of the region from the beginning of its settlement. As a result, older people and their households are frightened when they see cyclonic clouds in the sky or feel the light breeze over the rivers and the sea. Various testimonies suggest that geophysical and residential location are some of the major factors that contribute to increasing the vulnerability of older people and their households in the region. The location of the house is important in protecting the house from certain extreme events. However, most of the houses in the region are located very close to rivers where they have khas land [government wasteland where no one has ownership rights], which makes the houses vulnerable to extreme weather events. The extent to which older people are vulnerable to extreme weather events may thus depend on where they live and how they are settled (Alam & Collins, 2010).

4.3 Low-quality Housing and Lack of Living Amenities

The vulnerability of older people in the coastal region is also related to their housing quality and available facilities. A sturdy house can protect them from the strong winds during cyclones and floods. In the study area, most live in houses made of corrugated iron and wooden pillars or bamboo posts. They could not build their houses with bricks or other sturdy materials due to financial constraints. The houses are all closed and are congested. Older people often live in unhealthy, unsafe and unsuitable shelters. Chandresh, a 77-year-old widower, lives in Haritana village. When asked if his house is made of bricks, he replied:

> Bricks? How can I build my house with bricks? It is made of tin, which OXFAM [an NGO] provided after Cyclone SIDR in 2007. It was a low-quality tin that had many holes and defects. It could fall in the next cyclones. Rainwater leaks through the tin sheets when it rains.

It was found that the structure and foundation of the houses were very weak, as the materials were not strong, making the elderly and their households increasingly vulnerable to extreme weather events. Shuvasm from Haritana village reported on the condition of his house:

> Tin and wood are the main materials in my house. The government gave it to me after the Cyclone SIDR. However, it is rusted, has small holes and has been damaged [over time]. During the rainy season, rainwater enters the house. It is a normal house. It is not safe to protect us from cyclones and floods.

Most of the houses were donated by non-governmental organisations (NGOs) or local governments in the aftermath of Cyclone SIDR in 2007 or later, when the houses collapsed. Galvanised tinplate was used in the construction of most houses, which are located close to the river where they own pieces of land. This places the house in a vulnerable position. Abul, a 70-year-old casual boatman from Padma village, said:

> The house is made of corrugated metal sheets. I got this tin sheet as a relief after cyclone SIDR in 2007; Songram [an NGO] provided it. Now it is not in good condition. It has become rusty.

He also has no bed or cot. They have some Kantha (a type of embroidery used in eastern South Asia, especially in Bangladesh and the neighbouring Indian states of West Bengal and Odisha, as an alternative to bed sheets and quilts), but necessary clothing is scarce. The toilet he used to have collapsed during cyclones, so he rebuilt it on his own. Even though the house is usually safe, he believes it could collapse again if there were another cyclone. Most people in the area do not have beds. Chandresh, one of the oldest male respondents, said of his housing situation and amenities:

> … No. Where will I get the bed from? I do not have a bed or bedding. I lie down on a sheet made of my son's wife's discarded clothes. I do not have enough winter clothes. I have a thin blanket and a coat that I got from the government as relief. It is now torn because of overuse. It is very difficult to bear the cold winds of winter. I stay close to my grandchildren at night so we can keep each other warm.

Most older people have a small cot (locally called khat), which is usually homemade from local wooden materials, such as wood from old boats. Rizeeya, a widow from Badurtala village, said:

> I have a broken bed frame. I have tied it tightly. I have no quilt and no Kantha. I have a toilet. It was provided by an NGO after Cyclone SIDR … . My house is not safe during cyclones and floods. It is still broken.

However, most of them have toilets that were provided by NGOs after the Cyclone SIDR in 2007, but their condition is not good for use. Chandresh said, "I have a toilet, but it is ruined. I got it from a sanitation programme after the SIDR in 2007 and it is only the toilet for my family members."

Some have access to electricity but cannot pay the bill regularly. One of the male respondents from Haritana village said:

> … Yes. My house is connected to the electricity grid, but I cannot pay the bill. The bill has been due for six months. The electricity company told me last week that they would cut the electricity supply if I did not pay the bill (Chandresh).

Similar comments were made by Rashidul, who said, " … [although] electricity is available in my house, it costs me more. […] I used to use a lamp and a lantern." However, some older people, for example, Fakir Shah, reported that they [his family] did not have access to electricity. Some depend on the solar system they have installed and pay the cost in monthly instalments. Afzal Hossain has not yet received electricity. He gets electricity for lighting from his children who have installed a solar system which cost them about BDT30,000. Still others, such as Motaleb from Charlathimara village, live in small houses provided by the government, which are particularly vulnerable to extreme weather events due to their location near the river, age and structure. It is evident that poor housing quality and lack of housing facilities can contribute to the vulnerability of residents, especially older ones, to extreme weather events.

4.4 Poverty

Poverty is probably the most important factor in making older people vulnerable to extreme weather in coastal areas. Poverty is linked to a lack of access to resources (land, house, food), employment, social support and health care, which affect the lives of older people. Another universal aspect of poverty that makes it particularly painful and difficult to overcome is vulnerability. The frequent extreme weather events in the region have a greater impact on older people due to their poorer financial situation compared to the general population. Older people are more vulnerable than others after a natural disaster due to a lack of financial resources (Chau et al., 2014; Fees et al., 1999; Healy, 2003). Most older people in coastal Bangladesh lived below the poverty line; some had no income, while others made a living by begging and with the help of informal community and family support. Liakat, an 80-year-old who claimed that he was a freedom fighter, lived by begging. He did not have enough strength to work. He said, "I cannot do anything anymore. I live by begging. I eat what I get from people - rice and money."

The ability of older people to earn an income depends on their physical ability and access to and availability of regular work. However, most older people in the study area and their families do not have a regular source of income. For example, the income of Palkey Begum, a widow from Ruhita village, is not secured through her daily work. She could earn BDT150 a day if she had work, but she does not get regular work. Alfaz can earn 200 to 500 BDT per day if he can work. However, he was not always able to find work to earn an income. Then it becomes very difficult for him to meet his daily needs. Abdullah, an 80-year-old man, said:

> I do not have regular work anymore. I earn 10 to 30 Taka from some odd jobs I do every day. And I get five hundred Taka per month as Old Age Allowance [from the government]. [...] It is difficult to maintain my family. I take the essential foodstuffs from the grocery on credit. I have not yet managed to pay the due from last year. Three of my ducks lay eggs, and I sell them for 40 Taka per Haley (four pieces).

Further, older people are not paid enough for their work due to age and restricted ability. Some older people said that because of their old age and physical inability to do certain jobs, they are paid too little compared to young people. They believe that this is one of the factors that is gradually pushing them into poverty. A 70-year-old from Padma village, who lives with his son, said, "Sometimes I earn some money [when I can work]. But the young workers get five hundred Taka as wages, I, as an old man, get two or three hundred Taka for my work."

Thus, joblessness and sometimes underpayment often drive them into poverty, which makes them financially vulnerable. The financial crisis sometimes makes older people in the region vulnerable to borrowing. Older people who are in financial difficulties and have no close family members or friends have to borrow from others. Shuvasm from Haritana village, 60 years old and a small vendor, cannot support his family of four with his daily income. He sometimes borrows from people when he needs and pays them back later. Momina, a poor widow who lives with her younger son and daughter at the dam near Padma village, however, can neither get help nor borrow from others. She said: "My income is sometimes 100 takas and then again only 50 takas. [...] What will I do if it is not enough to maintain my family? Who am I going to tell? ".

Older people cannot afford to eat nutritious food in the area because of their poverty. They have to wait for someone to give them this kind of food, or they visit a relative's house. An 80-year-old man, Rashidul, explained:

> Sometimes a glass of milk is out of the question for me. Fruits are not usually eaten. When I walk into the market, I look at the fruits but cannot afford them. If I go to someone's house or if someone gives me something, I can eat. For the last five years, I have not been able to buy any fruit. I have some fruit trees, but they were damaged by cyclones at different times. I do not see the fruits; how can I eat them?

These accounts show that the elderly living in the area have little access to adequate food and nutrients due to their poverty, which makes them vulnerable to poor health and various diseases. The older people in the area most likely suffered from chronic income poverty in their childhood because their parents were also poor and worked irregularly in their later life. One of the participants shared his experience of poverty:

> I have had a bitter experience of scarcity. When we were very young, I used to pick paddy from Char [Island] with my younger sister. We boiled the paddy, dried it in the sun and husked it with demki [local wooden bowls for rice preparation] to make rice. Then we cooked the rice to eat. (Alamgir, 62 years of age, male, carpenter from Badurtala village).

Despite his efforts, Alamgir still remained in a state of chronic poverty, living in precarious socio-economic conditions as a casual carpenter. Naima, a 68-year-old widow from Padma village, also suffered from poverty since childhood. Due to her father's inability to pay for schooling, she had to work from a young age. Her father did not give them enough to eat. To eat rice, they had to collect the fallen paddy from the rice field after the crop was harvested. Despite her hard work, she claimed she could not change her fate. She lived in almost the same socio-economic conditions as she was in her childhood. To survive, she has to work every single day. Her son cannot help her either. She said that her son lives separately from her. He has two children and has huge expenses to support his family. She described:

> ... Sir, my family cannot be fed with the income I earn. I have to buy oil, salt, pepper, rice and some betel leaves. I have to buy some shells for my goats. I also have a daughter who needs soap. With the two hundred Taka I earn, I cannot cover [my expenses].

Poverty is a major risk factor when facing an extreme weather event or disaster (Yoon, 2012). As poverty is very high among the older people in the region, it is reasonable to assume that they were particularly vulnerable to disasters due to financial difficulties. Chau et al. (2014) argued that poverty limits the ability of older people to acquire the resources necessary to maintain their wellbeing. For example, older people living in poverty cannot afford adequate food and may not have access to basic disaster recovery resources and services such as shelter, clothing, transport and food, making them vulnerable to extreme events.

4.5 Informal and Precarious Work of Older People

Older people are most affected by unemployment following natural disasters like cyclones and floods. They are likely to be unemployed for longer and spend more time looking for work than young people in the region. They often face discrimination when it comes to wages and employment (Weller, 2021). Most older people in the region are engaged in traditional occupations such as farmers and fishermen. However, some older people are forced to change their employment year after year due to the impact of extreme climatic weather events. They work as casual labourers in small-scale enterprises, fishing and engaged in marginal farming, where they do not earn a regular income. Shuvasm explained:

> I sell betel leaves. I buy the betel leaves from the wholesale market and sell them door-to-door in the village. I do this from morning till evening. I have been doing this work for more than 15 years. Before I started this, I worked as a labourer in others' houses or land, doing farm work, tree planting and other work.

Since the study area is a coastal region, most older people and their families in the different villages make their living from fishing. They catch fish in the river and sell to wholesalers and vendors in the market. They also work as casual workers on fishing boats. Shamiul is one of the fishermen who has been fishing in the river from a young age and selling it in the bazaar to earn a living. He said:

> I catch fish and sell it. I go to the river and the sea, catch fish and sell it [in the market]. I have been doing this since I was a child. But I also work as a labourer and sometimes go fishing with others.

Haulader from Padma village used to do a similar job, catching fish in the river, but now he is not physically weak to do this work. Some run small fishing businesses in the village markets. For example, Fakir Shah from Padma village has been involved in fishing since childhood to support his family. He used to catch fish himself in the river and the sea. Now he buys and sells fish in the market after suffering huge losses in his fishing business due to Cyclone SIDR in 2007. He explained, "Now I buy and sell fish in the market and sometimes in the village. I buy two or three kilos of fish and sell them in the market." Although fishing is their main occupation, most also have to do other work if necessary, following extreme weather events and economic crises. Abdul, a 70-year-old man from Padma village, is now a boatman. Although he was poor, he worked very hard since childhood to make a living. He said:

> Now I sail a boat in the sea to catch fish and prawns with a net. I do not do anything anymore. If I do not work and catch fish in the sea, I cannot maintain my family [expenses]. If I pull a net into the sea, I get some earnings to feed my family, otherwise, I have to endure a lot of suffering.

Some elderly people are marginal farmers who are particularly vulnerable to extreme weather events. They grow various crops on their farmland. They also do odd jobs in other fields to earn extra income to support their families. Chandra, a 73-year-old man from Charlathimara village, is one of them. He also does various odd jobs apart from his regular farm work. He said:

> I am a farmer, and I do farming - I plough, harvest paddy and other things. [...] Sometimes I also work as a labourer [casual worker]. I have worked in people's houses with a salary from the beginning [in my childhood]. I also pulled the net in the sea to catch fish.

Some work as irregular casual workers, whose earnings depend on their daily work. These older people are more vulnerable than those who have regular jobs and businesses. Some of them work in other people's houses in the village, and others work in the fields and help the owners with managing their lands and agricultural activities. Some older fishermen have been catching fish in the sea and rivers for years, but as one needs more physical strength to pull a fishing net, they sometimes give up. Alfaz Hossain works as a day labourer. Occasionally, he pulls nets in the sea and river and works for people in their houses or fields. However, he cannot stay in the river for long because he feels dizzy after a while due to physical weakness.

Alamgir now works as a carpenter, although he used to pull nets and catch fish in the sea with trawls and drag boats in the middle of his working life. He recounted a sad experience from his fishing days:

> Once pirates captured my boatmen and demanded 300,000 Taka as ransom. I paid and released them. Then I bought crab nets for 150,000 [150k]. The Coast Guard set them on fire. Now I work as a carpenter. I practised that with my father.

These kinds of experiences - assault and robbery of fishing boats in the deep sea and seizure and burning of fishing nets by authorities for illegal fishing - are not uncommon for older people and their households in the fishing community, forcing them into financial hardship. In some cases, once older workers leave formal employment, they find it difficult to adopt another employment for the rest of their lives. For example, Muhibul, a 79-year-old uncertified freedom fighter and retired police officer, runs a tea stall on the riverbank near his home in Badurtala Clustered Village.

It is even more difficult for older women, especially widows, to find work. Most older women interviewed said that sometimes they have to work as casual workers in others' houses or catch fish in the river as a helping hand with their fellow men. Sometimes they participate in the 'Food or Cash for Work' programmes run by the government through local authorities to repair roads and dams. Momina, a 60-year-old widow from Padma village, also works as a day labourer [casual labourer] like Hossain. She said:

> ... I occasionally catch fish in the [wetland] of Char [Island] and sell them. [...] I sell them to earn a living. Before, I worked in the local police camp; I dug earth to fill up roads [repair roads].

Some older women were small shopkeepers or tea sellers. They sell tea throughout the day to earn some income that helps them pay for daily expenses for their family. Rajina from Ruhita village works with her husband as a tea seller in Ruhita Bazar at the dam. Naima, a 68-year-old widow from Padma village, worked as a casual labourer, sometimes as a fish dryer and at other times as a fish catcher. She said:

> Sir, now I work daily as a fish dryer [work of drying fish] in another person's fish factory. People dry the fish there where I work as a labourer. I go there in the morning and come back just before dusk.

In addition, many believe that their lack of physical strength prevents them from receiving the same employment and wage opportunities as other adult members of society. They are also underpaid. In most cases, employers do not hire them because of their age, as they cannot do certain jobs well. Some employers believe that older people cannot do the work assigned to them quickly, punctually and properly. Employers, in turn, want to offer lower wages because of older workers' reduced ability to work. Rashidul, an 80-year-old participant from West Haritana village, said:

> Yes. People do not want to give me a job because I am very old. They say that such an old man like me is not able to do the work. They also pay me very little compared to others."

Over time, their diminished work abilities and underpayment led older people in the community to believe that they were no longer needed in the community.

4.6 Lack of Land and Resources

Land and other resources can help reduce the vulnerability of older people and their households during and after extreme weather events. Stocks of assets and resources help reduce vulnerability as they can be converted into income, food or other basic needs (Moser, 1998). However, scarcity of land and resources is one of the most important determinants of vulnerability among older people in the region. In most cases, they did not have cultivable or residential land and other assets. In the study region, ownership of ancestral land and assets among descendants had decreased over time. This was due to the continued sale, losses from natural disasters such as erosion of riverbanks, and subdivision among descendants over time, gradually making them vulnerable to extreme weather events. Many reported that their grandparents or parents owned some farmland but sold it in the past to support their family. They even had to sell what they had received from their parents and grandparents, especially after extreme weather events in the past, to raise their children. For instance, Chandresh from Haritana village does not have much land. The land he had received from his parents had to be sold at different times due to socio-economic circumstances and extreme weather events. Chandresh has three sons and a daughter. He sold some of his lands; he gave away some of his lands to his sons and sold some when his daughter got married.

> I do not have any cultivable land. I sold all the cultivable land I got from my father to support the family's livelihood. However, I have 10 decimals of residential land which I have left for my three sons to live on. My grandfather owned about 14 acres of land, of which my father got about 4.5 acres, but he sold most of his cultivated land. I have no other resources. [...] My son, with whom I live, also has no other resources.

This region often experiences economic crises due to extreme weather events, which force farmers to sell their land, leaving them even more vulnerable to the next extreme event. Some did not own any land, either residential or agricultural. Despite being illegal, they live on government land or a dam. Shamiul, an 80-year-old man from West Haritana village, is one of them who lives with his family on government khas land. He said:

> I do not have a single piece of land, no residence. I live on khas land [unproductive land owned by the government]. My father had some land. He survived by selling it. [...] We could survive by farming if there were some lands. [...] I have no assets. I have no trees, no business, nothing.

Landlessness means loss of livelihood, sometimes homelessness or inadequate shelter, loss of identity and displacement of family, which increases their vulnerability (Reale & Handmer, 2011). Raizuddin, a 69-year-old respondent from Ruhita village, said he does not own much land. He owns six kathas (15 decimals) of land that he got from his father. He bought some land, which he distributed to his three sons and two daughters.

In the rural coastal area, many older people live with their family members, especially with their sons. In this study, most participants live with their sons even in old age and bequeath most of their land holdings to them. Rashidul from Haritana village said that his parents had received some land from his grandparents, which his parents later passed on to his brothers. He was given some land, which he in turn distributed to his sons. Now he only owns the residence where he lives with his wife. Some of the participants only own the plot where they live. They have no cultivable land. Shuvasm from Haritana village has no land to cultivate. He has only eight decimals of land that he got from his parents, where he now lives with his family. Apart from four ducks, he has no other assets. Chandra, a 77-years-old man lives in Charlathiara village with his wife and two sons and their spouses and children in a single residential house. He says he does not have enough land. He has about an acre of land for cultivation and this residential plot. He said:

> My grandfather had 6/7 acres of land. My father got almost one acre as his share. I do not have anything [other assets], sir. I have no cows or goats. No poultry. I used to have some. I sold everything to pay for the treatment of my younger son who is suffering from a liver problem.

Most respondents received little or no land assets from their parents. Alamgir Hossain and Fakir Shah are among the respondents who also have no land. Hossain's mother has ten kathas [about 33 decimals] of land and his two brothers live there. His father had two kuras [1 kura = 66 decimal] of land. He lost his land after his second marriage. His father did not get land from his grandfather. He got the land from one of his uncles who owned a huge amount of land, and he gave his father two kuras of land. Similarly, Fakir Shah said:

> I have no land and no residence of my own. I live next to the dam of the Water and Power Development Authority (WAPDA). My father had no land. My grandfather had some land, but he sold it. My father got a little of it. He got 35/40 decimal (12/13 Katha) land from my grandfather, which he sold and fed us. I have two goats.

As presented above, older people are more vulnerable to the effects of cyclones and floods because they have less land and fewer resources. It is said that older people who own land derive power from it, which puts them in a better position than those without similar power resources. They are then unable to maintain themselves and become more dependent and vulnerable, which reduces their status in the family and community (Salamon & Lockhart, 1980). Access to land and property, which is also cited as an indicator of vulnerability in numerous studies on vulnerability, can enable older people and their households to respond quickly to the impacts of extreme weather events through leasing and mortgaging. However, the main concern of the landless is about their livelihood strategies. Older people who have cultivable land can take precautionary measures for the next extreme event, while the landless elderly and their families are less able to do so. Landowners have key assets to withstand extreme events, which is not possible for the landless, thereby increasing their vulnerability (Paul & Routray, 2011). Older landowners also have more social connections than the landless. The consequences for the landless group are thus devastating. This process is accompanied by sociocultural deprivation and dislocation, which play an important role in determining the intensity of indirect disaster impacts and the vulnerability of older people in the area (Haque, 1997).

4.7 Dependence on Children

The dependency of older people on children is on an upward trend and is expected to triple in the next few decades due to limited access to social protection programmes and financial insecurity in old age (BBS, 2015). In Bangladesh coastal regions, older people depend on others in their family or community to meet their basic needs during and after extreme weather events. Dependence on family members, especially sons or daughters, is one of the vulnerable conditions for older people in the region. Most older people reported that they need to depend on their sons or daughters in some way for their daily needs. Chandresh, a 77-year-old man, recently lost his wife. He feels lonely and needs the help of other family members to make a living. His older sons are not much better off. Therefore, he has to depend on his younger son. He said:

> I lost my spouse a few months ago and now live alone. My three sons and three daughters live separately. [...] I live with my younger son who has two children - a son and a daughter - and his wife.

But he faces some problems living with his son's family. He depends on his son for meals and other things. Sometimes he has to bear the rude behaviour of his son's wife. He said, "[...] Yes. I have had some difficulties living with my son [and his family], but I am afraid to say anything because if I do, they will throw me out of the house. If I say something, they tell me to go live with others".

Others living with their children also reported similar living conditions. Amena Begum, 80 years old, cannot work. It is hard for her to live alone. She shares her house with one of her daughters who suffers from some disabilities. She depends almost entirely on her daughter, who lives with her, for her household and personal needs. Her daughter does all her work as she cannot work for herself. She said:

> [...] My son's family, who live in another place, lacks money. He cannot feed his own family. What will he do for me? [My] daughters help me in the crisis. My five daughters give me clothes, money and meals when I need them.

Debnath lives with his wife, two sons and their wives, and grandchildren. His daughters and sons have married. Although he is a casual carpenter, he sometimes has to depend on his sons when he cannot go to work to earn money due to his vulnerable health condition or unfavourable weather. Debnath said he cannot go to work as he used to be able to, so he is no longer able to provide adequately for his family. As for contributing to family expenses, he cannot tolerate his sons' uncooperative behaviour. He sometimes feels bad and does not bother to go to his sons' shops to fetch money. He realises that his sons consider him a burden:

> Family members, I think my sons consider me a burden in the family. I cannot take up any more work. Therefore, I have no income. I cannot support myself. Therefore, many consider me a burden. My sons do not ask me before going to do any work.

Almost all older respondents feel that they are viewed as a burden by their family members, their children or grandchildren with whom they live. They believe that every son or daughter has to take care of their children/family members. However, they find that their sons' or daughters' income is not enough to support their parents after taking care of their children and family members. Therefore, their parents and older relatives living with them are considered a burden. A 79-year-old Muhibul from Badurtala village said:

> They [sons/daughters] also see us [parents] as a burden. Every son/daughter has a family of their own. Take care of their immediate family but consider it a burden to take care of their parents.

Sakimun, a widow from Haritana village, had a similar experience. When asked if her life was a burden for her or her family members, she remained silent with tears in her eyes. She said that her sons and daughters sometimes asked her how she was coping with the basic necessities, but whenever she wanted something from them, they said they were in crisis. Then they would say, "We are in trouble, how can we give you anything?" Ismail Hossain, a 75-year-old poor farmer, lives with his wife in Haritana village. Although he has some land, he cannot farm it regularly. Further, he is suffering from various diseases. But his daughters and sons do not take care of him and his wife. He expressed it thus:

> My sons and daughters do not provide meals. I am considered a burden by them. These are all hazardous [for me] at this moment. They [my sons and daughters] do not want to ask us if we have eaten or if we need to drink tea, if we need to eat warm rice or if we need to wash our clothes. They think it would be better for them if we died. What will you do if I tell you about my suffering?

He claimed that many people are of a similar age to him, but their sons do everything to please their parents. There are others who do not take care of their parents. He claimed that his sons and their wives did not help him with everyday needs, such as washing clothes. But when it hurts their interests, they use swear words and reprimand us, he said.

Despite the problems in their relationships with their adult children, older people in the area still maintain some relationships with them because they have no other means of making a living. It appears that older women are more vulnerable than their male counterparts due to local cultural and religious issues. They face some problems, but they sort them out and live together. When asked if she has problems living with her daughter, Rizeeya, a widow from Badurtala village, replied:

> Sometimes I have problems living together with my daughter. My daughter is unemployed. She cannot support herself. How is she supposed to support me? My daughter's house is right next to mine. My daughter would give me food if I could not cook. Sometimes there are disagreements with my daughter. But she would call me later. I do not keep it in mind.

Naima, a widow over 60, lives with her daughter and son-in-law near her son's house. She does not live with her sons because of various family problems. She said that she had some problems with her son and daughter-in-law when it came to taking care of themselves. Naima believes that her family members - her sons, daughter-in-law and grandchildren - should take care of her when she is old. They should cook for her and serve her food. They should buy the things that she needs. But she has to work to survive in her old age. This is unacceptable to her. Some respondents have no respect for their lives because of their socioeconomic problems, and the unwelcome behaviour of their family members. In their opinion, it is better to die than to live dishonoured and with constant problems. Momina, for instance, lost her husband many years ago and had to endure a lot of suffering while raising her children. She said, "I pray to Allah to take me to the hereafter while keeping my body strong and free from pain." This is very similar to Fakir Shah's: "Many times, I have prayed to Allah in sorrow, asking Him to grant me death so that I am not in these circumstances." This happens especially when there is economic insolvency. Various statements indicate that the older people in the area continue to live in inadequate living conditions and are unable to make a living because they do not receive adequate family support and self-sufficiency.

4.8 Limited Access to Food and Nutrition

Limited access to food and nutrition has detrimental effects on the health and well-being of older people, which in turn can limit their ability to work, engage in daily activities and live independently, which can limit their capacity to endure and cope with the impacts of extreme weather events (FRAC, 2019). The lack of cultivable land and resources affects the availability of food and nutrition for older people in the region, as they have less food available due to lower production capacity. Most older people were found to suffer from malnutrition because they did not have the necessary food. Some of them find it difficult to get food. Chandresh, a 77-year-old widower from Haritana village, claimed:

> [...] I have to face various problems to get food. There is no other way but to bear these difficulties. There are some problems that I cannot tell people about. My son's wife does not give me enough food. She keeps the food at a safe distance so that I cannot reach it. When they cook fish, they eat most of it and give me a small piece from the tail side. I never get the head of the fish. They eat it [the head of the fish], keeping it for days. They bring other food like biscuits from the market and eat them; I just watch but do not eat.

Many older people depend on their family members, especially their children, for their daily food and drink, as they are unable to earn and manage their food before and after extreme events. The food intake of older people depends on the availability and accessibility of food in the family. For some older people in the region, it is difficult to obtain their daily food. Older people living with their sons' or daughters' family members get less food and fewer nutritious items. This is because their sons and daughters first think about their own children rather than their parents. Most in the area have little or no access to nutritious food such as meat, milk, fish, eggs and fruits. Chandresh said again:

> I eat some of it [nutritious food], but it is rare. Where shall I get money to buy these foods? The price of a dozen eggs is BDT 100 and a KG of milk costs BDT 60. The price of meat is high for us; a kilogram costs BDT 500 to 700. [...] Most of the time I cannot eat nutritious food. If I buy this kind of food, I cannot buy rice or curry dishes. I usually eat meat once a month when we can earn more. We cannot eat fruits, but I can eat coconuts as I have some coconut trees.

Similarly, Fakir Shah from Padma village and Shuvasm from Haritana village in the Union can hardly get nutritious food because of their poverty. They said that they have to wait for months or even years to get certain food like meat, milk or fruits. Most usually eat rice, curries, potatoes, vegetables, dal, and bread. Rashidul from Haritana village eats a small amount of food at different times of the day. He eats rice only at night. During the day, he takes some tea, biscuits and parched rice. However, due to his lack of financial resources, he cannot get nutritious food. Eighty-year-old Abdullah has meals twice a day - at noon and in the afternoon. In the morning, he drinks water and has a small amount of tea. Sometimes he also eats rice and bread with curry in the morning. "Two to four eggs are eaten a month, fruit and meat are not eaten. How can I eat and with what?" - he said. He eats fish when someone gives him a few pieces or when he can catch fish. Alfaz believes he needs nutritious food for his body, but he cannot buy fish, milk, meat or eggs.

Older people who are vulnerable to dietary stress have a lower physical reserve, which limits their ability to recover vigorously in the face of an acute health threat or stressor from extreme events (Porter Starr et al., 2015). Some older women who have lost their spouses find it more difficult to get enough food and nutrients. "I do not remember eating any fruits after my husband died, " said Palkey, a widow from Ruhita village. She does not have milk, eggs and other nutritious food. Momina, a widow from Padma village, has to struggle to meet the daily food needs of her children, a son and a daughter, with whom she lives. She said:

> [...] Sir, I do not even see this food [nutritious food] with my own eyes. I think it has been a year since I ate meat. Fish is eaten a little when my son brings it. Milk, eggs and fruits are not eaten. Where would I get them from, sir? Don't I need money to buy them?

Most of the older women said they eat the standard food during the day - potatoes, dals, rice and vegetables. Fish, meat, eggs and fruits are beyond their means. Some only eat eggs from their chickens or ducks. When asked how many times a day you eat, Rejima, a 65-year-old woman from Padma village, replied:

> How often will I eat? I suffer a lot as I eat only once a day. I cannot eat now. Fish, I can eat if it is brought by someone, otherwise not. Meat, especially beef, is not eaten except at sacrificial feasts. [...] People sacrifice cattle. They give me some meat and I eat it.

What about milk and eggs? "I can eat a little egg. Milk, I cannot take at all. I cannot buy any other food if I buy milk." Naima, a 68-year-old woman from Padma village, said she did not cook today because she was under pressure at work. When asked, "When did you last eat meat?" she replied:

> I cannot say exactly when I ate meat; I cannot remember, but it has been many years. I can't afford milk and fruit. Where am I supposed to get these foods from?

The experiences of older people in coastal Bangladesh provide holistic insights into access to food and nutrition in the region. Most eat rice once to three times a day, sometimes a piece of bread or biscuits. They cannot have nutritious food because they can't afford. In addition, it can be more difficult for older people to eat healthy food because of their age and financial difficulties. This can reduce their physical reserves and limit their ability to recover in the face of a serious health problem or stressor. The physiological and sociocultural factors that affect food intake and behaviour are also important, and older men may be more vulnerable to malnutrition (Porter Starr et al., 2015; Thompson et al., 2017). Food insecurity harms the ability of older people and their family members to cope with natural hazards and crises. Hunger and food insecurity have an adverse impact on disaster risk because they significantly increase the vulnerability of the elderly population to extreme natural hazards (Garschagen & Schmidt, 2015).

4.9 Health and Illness

The most common diseases affecting older people in Bangladesh include cardiovascular dysfunction, diabetes, arthritis, osteoporosis, skeletal deformities, sleep disorders and dementia. People over 50 often suffer from cancers such as prostate, and lung, stomach, colon, breast and cervical cancers. More than 50% of people over 65 in rural and coastal areas suffer from arthritis, which can cause pain and lower quality of life (Islam & Rahman, 2017; WHO, 2018). Consequently, they struggle to be strong both physically and mentally in the face of challenging socio-economic and extreme weather conditions. The older people in the study area find it difficult to move from one place to another. Some could not even go to a nearby café or grocery shop. As 77 years old widower, Chandresh, narrated:

> I used to walk to the shop near my house once in the morning and once in the evening and have tea there.' [...] I used to participate in various cultural activities but now I cannot go because I cannot move in winter and summer.

Physical weakness is one of the vulnerabilities of older people living in the region. Some stated that due to the decline in their physical strength, they are no longer able to perform daily activities such as bathing, washing clothes, etc. without the help of other family members. They are also no longer have the capability to participate in various social programmes as they used to. For example, Rashidul from Haritana village can no longer do his daily activities without the help of others. He said:

> Normally, I can go about my work, but when I take a bath, I have problems; I cannot bathe in the pond. I cannot wash my clothes. My wife helps me to get water for a bath and to wash my clothes.

He also said that now he cannot participate as actively in social life as he used to. Sometimes he goes to religious ceremonies, but not to cultural activities or programmes. Again, chronic health problems are one of the most vulnerable conditions for older people in the region. Many of them suffer from various critical health conditions such as cardiovascular diseases, kidney diseases, digestive disorders, stomach problems and chronic pain. For many, their adult children cannot afford the medicines prescribed by the doctor to cure their diseases. Alfaz, a 66-year-old man suffering from pain, said, "I have many problems in my body now. I have pain all over my body now. I cannot do any work. I cannot carry the load. I do not have money, so I cannot take medicine."

Arthritis and body pain are common health problems among older people, hindering daily physical activities and increasing mental distress. Raizuddin, a 69-year-old man, also reported:

> I am no longer hale and hearty. I am suffering from various diseases. I have been suffering from acute headaches for about two to three years. I have to take tablets [medicines] daily - one red and one white. These two tablets cost me six Taka.

Older women, especially widows who have to work longer hours to support their families, are more likely to suffer from physical and mental problems. Palkey Begum, a widow from Ruhita village, has been suffering from various diseases for a long time because she was not treated properly after an accident. She said:

> Sir, there is a problem with my body. I am suffering from many diseases. One day, trees fell on my head and wrist while I was picking eggplant in the vegetable garden behind the bamboo bushes. Some people who were cutting trees did not notice me. They cut the trees and dropped them on the land .... One branch of the tree fell on my head and another on my wrist. I was seriously injured. I have acute pain in my wrist now. I cannot do much work. Again, I was injured during Cyclone SIDR in 2007. I have to take medication.

Rajina, a 63-year-old tea vendor from Ruhita Battala Bazar, like Raizuddin of Ruhita village, suffers from poor health with pain in various parts of her body. She said, "I am in poor health. I have pain in my waist, knees, wrists and chest. I also have a bone problem." Sakimun, a 64-year-old woman from Haritana village, also suffers from chronic pain in her wrists and waist. Her physical problems make it impossible for her to do any kind of work. She said:

> ... [Oh] my son, I am burdened with so many diseases. I am on the verge of fainting because I have pain in my waist. My stomach feels like it is going to intertwine with my waist. Then I could not hold still anymore. My neighbour's housewife used to take me to the talim [in-house religious gathering] by holding my hand and helping me return home.

Researchers have established that deteriorating physical and mental health, frequent falls and limited care can make older people vulnerable (Barbosa et al., 2019; Brocklehurst & Laurenson, 2008; Sarvimäki & Stenbock-Hult, 2016). To build adaptive capacity to cope with socioeconomic hardship and extreme climate change, older adults need to have the physical and mental strength and understand their limitations. However, the current study shows that poor physical strength and reduced mental capacity of older people increase their vulnerability to various extreme events. Their health status may lead to the need for evacuation in emergencies during extreme weather events. The lack of basic health services and limited access to health facilities also increase the vulnerability of older people in the region (Malak et al., 2020).

4.10 Relationships and Social Network

Social networks foster interpersonal relationships between individuals and groups in the community. People can access social support, material resources and various forms of capital (cultural, economic and social) through social networks and support (Andrew, 2010; Bourdieu, 2018). The extent of older people's vulnerability to extreme weather events sometimes depends on their social networks, relationships, which help them through the crisis, and their adaptive capacity. However, in coastal Bangladesh, people have suffered significant losses of crops and livestock over the years, leading to a loss of traditions and a weakening of social cohesion and family ties (Khan et al., 2014). Some older people indicated that although they have good relationships with their neighbours and family members, they are less able to maintain social networks due to physical incapacity.

They sometimes participate in religious activities and programmes when possible but cannot participate in all social and cultural programmes in the community. However, they try to maintain some close relationships and regular communication with their children who live outside their local community. Abdullah, an 80-year-old resident of Charlathimara village, said:

> If there is a programme in our village, I usually do not go. I only go if there is a special religious occasion. I communicate with my younger son who comes once a year, and I often talk to him. My relatives are no better off than I am. None of them has land and they live very close to me.

Social networks and other support systems can reduce vulnerability among older people by providing material and in-kind support and advocacy (Kreager, 2003; Schröder-Butterfill & Marianti, 2006). However, weak social networks can increase the vulnerability of older people and their households if their support system is not strong. Economic insolvency has been found to have some social connectivity among older people and their households, but support systems are not strong enough to reduce vulnerability during socio-economic crises and natural disasters. Razeeya, a 70-year-old widow from Purbo Badurltala village in the Union, shared her experiences:

> Relatives are poor. No one can give me anything. They cannot take care of themselves. [...] The neighbours cannot help each other. If I can do something for them [for other people], they give me a little, and if I cannot do anything, they give me nothing.

Nevertheless, some older people do not participate in sociocultural activities during the day because of their workload. They work all day and rest at night. Shuvasm, a participant from Haritana village, stated:

> I do not participate in any social programmes, but sometimes I participate in religious activities. I go to work in the morning and after I come back from work sleep in the evening. I came here because you called me, otherwise, I would not have come. Normally, I do not go out. After work, I have my dinner and go to bed.

Fakir Shah, a small fish seller from Padma village, lives in the same conditions as Shuvasm. He is one of the most cyclone-affected members in the community and lost everything in the last natural disaster. He is busy buying and selling fish in the market all day. He said:

> When I get up early, I have to catch fish from the nets. After I buy fish, I go to the bazaar. When I return, I go to catch fish again. I spend a little time with others in the bazaar. I cannot take a rest. If I take a break, I have to starve the next day. I have no one to help me with my work.

The constant threats of extreme weather events throughout the year make them vulnerable to poverty. As a result, they have to spend much of their time each day looking for work and food, which affects their social connection. Some do not have many friends or relatives in their community because most of them have died of old age.

Because of their involvement in various households and odd jobs for livelihood, some older people, especially widows, do not have much time to participate in local programmes. Their social relationships are also relatively weak. Momina, a widow, does not have time to participate in social events in the area because of her daily work for a livelihood and household chores. She focuses more on work than on social events. Having lived in poverty for many years, she knows the links between socioeconomic status and family ties. She thought:

> The relationship with kinship largely depends on economic status ... Brothers help their sisters if they are relatively well-off, otherwise, it is not possible. [My] brothers live on [nearby] Char-[Island]. Twelve years have passed since I last visited them. I meet them on the street, but they do not come to [my] house. Nor do they ask about my woes and wellbeing.

Older people like Shuvasm, Momina and Fakir Shah, for example, may lose their social network due to their work commitment and may not be able to maintain regular relationships with other members of their community. It is evident that having a good network of family and friends and living in a good neighbourhood can increase older people's standard of living and reduce their vulnerability. In general, in the coastal region of Bangladesh, older people have some good relationships with their children and family, but not as good as they would like with their neighbours and support organisations. Older people with low confidence in social networks, knowledge sharing, solidarity and cooperation were more vulnerable than those with high levels of social capital, as evidenced by, for example, participation in mutual aid, social cohesion or other collective activities. Older people who had not established social networks and bonding relationships early in their lives were likely to be excluded or exempt from informal vertical and horizontal social networks that provide for the family, extended kin, neighbourhood, community groups or peer groups. The interaction between low social capital and the resulting need for care is illustrated by older people who are dependent on care and assistance and have 'no carer' (Eeuwijk, 2006).

4.11 Discussion and Conclusion

The vulnerability of older people can be considered a multifaceted construct where environmental influences such as culture, economics and politics interact with biological mechanisms throughout their lives. This chapter discusses the sources of vulnerability of older people in coastal Bangladesh. The vulnerability of older people is due to poverty, poor health, inadequate housing, nutrition and support, and insufficient employment and income opportunities (HelpAge International, 2013; Islam & Khan, 2018). The experiences of older people living in the coastal region of Bangladesh showed that the vulnerability of older people begins with their settlement in the vulnerable area. Older people in the area have been vulnerable for more than 150 years since their grandparents or forefathers started clearing the jungle in the wetlands and settled here. Their ancestors had to fight against nature and extreme natural calamities from the beginning of their settlement in this area. This vulnerability was passed down from generation to generation without any permanent solutions. Most older people's settlements were located near the river and sea, which made them more vulnerable to extreme weather events.

The vulnerability of their settlements to extreme weather events, especially cyclones and floods, was related to three mutually reinforcing factors: location, the likelihood of an extreme climatic event occurring, and the extent to which their structures can be damaged. Their houses were considered vulnerable because they were built with cheap and poor building materials such as mud, corrugated iron sheets and ridged tins. Many of these houses were light, typically wooden frame structures, which made them particularly vulnerable to cyclones. Building houses that were resistant to cyclones was very difficult for older people, the disaster victims, due to acute poverty and lack of resources. They could not afford the materials such as sand, bricks and cement needed to build cyclone-resistant houses. They also had limited access to local engineers who could build cyclone-resistant houses (Islam et al., 2018). Many of the houses in which older people and their households lived were of very poor quality. Most houses were provided by NGOs or local government organisations following extreme weather events. In some cases, the houses were built by older people themselves with their hard-earned money or with loans. However, for most, it was very difficult to build houses that could withstand extreme weather events such as cyclones.

Older people living in absolute poverty lacked the means to deal with extreme weather events, which further increases their vulnerability. Poverty and disempowerment were linked to a lack of access to necessary resources. In addition, poor older people may be more vulnerable to extreme weather events due to their dependence on family and community resources (Adger, 1999). In coastal Bangladesh, older people suffered from a lack of necessary amenities in their homes. They did not have enough warm clothes for the cold weather, and they had limited access to toilets. In addition, there were no mobility aids for them; some of them relied on family members or other people to get around or do their daily activities. In addition, the livelihoods of older people from low-income families in coastal areas are associated with endemic poverty and their vulnerability to extreme weather events may increase due to economic difficulties (Islam, 2018). Vulnerability in old age also appeared to be strongly associated with low material resources and poverty (Crooks, 2009). In general, poverty was associated with a lack of resources or assets, making it more difficult for older people and their households to address threats when they occurred.

Older people in the coastal region of Bangladesh had very few employment opportunities. Some jobs are not accessible to older people because of their poor physical strength and limited mobility. As a result, many older people, especially older women, preferred to stay at home and do household chores. The physical limitations of older people may have prevented them from moving around independently due to age-related problems such as frailty or disability. In addition, their mobility was further limited by a lack of sufficient financial resources. Their weakened physical and financial conditions led them to become increasingly dependent on their family members for daily needs in old age (Malak et al., 2020).

Land was the most valuable asset for rural older people to survive with dignity in the community and there was a link between asset availability and vulnerability of later generations. In the study area, grandparents and parents of older people owned some properties, some of which they lost due to extreme weather events or other social or family crises. Some older people who had inherited property from their grandparents or parents were less vulnerable for a few years but had to sell it to survive extreme weather events, which pushed them into asset vulnerability. The parents of the majority of the elderly were themselves poor and owned little or no property and land, which hindered their better upbringing, education and greater job opportunities, which is one of the causes of their vulnerability in old age. Moser's (1998) approach to asset vulnerability is based on the premise that the more assets' people own, the less vulnerable they are, and the more their assets decrease, the greater their vulnerability becomes. Since vulnerability was largely related to a household or individual's assets and insurance strategies relative to risk, the poor are less resilient to shocks and therefore more likely to become vulnerable (Crooks, 2009).

The old-age dependency ratio in Bangladesh was 8% (BBS, 2015). The increasing dependency of the elderly was more than offset by the decreasing dependency of children, creating the problem of economic support for the elderly in the country. Most elderly in the study area were socio-economically dependent to some extent on their family members and the social support system, as they were vulnerable to extreme events due to their economic incapacity and lack of resources. Before and after extreme weather events, when the social support system was inadequate, they depended on their children for daily necessities such as food, clothing and medicines. The older people were particularly vulnerable during a cyclone due to their lower assets and dependence on young adult family members. In addition, the vulnerability of people living in the coastal belt was greater due to low income or dependence on their son or daughter, which limits their ability to cope with such events. Older people living in coastal Bangladesh were particularly vulnerable due to their lower mobility and economic and cultural dependence on their children (Malak et al., 2020).

Focusing on climate change and seasonality, Shams and Shohel (2016) analysed the existing threat to food security and livelihoods in coastal Bangladesh. Their study revealed that food shortages occur repeatedly during the sowing and harvesting seasons. While the food deficit remains high (68–95%) from mid-August to end-October, it reaches its lowest level (21%) in December. The present study thus finds that food insecurity or vulnerability is a common scenario among the elderly population in the region. The food intake of older people in the study area was influenced by their ability to buy food or the availability and accessibility of food within the family. What they were given was what they ate, and nothing special was prepared for them. There were days when they did not get anything to eat. In contrast, nutritious foods such as meat, fish, eggs and fruits were only taken when they were given by a close relative, a neighbour or at a sacrificial festival of the Islamic religion during the year. They could not buy the goods themselves because they could not afford them. In general, rice, wheat, pulses and vegetables were the most commonly consumed foods, while fruits and sweets were the least consumed by older people. In addition, older people could not feed themselves regularly and their families essentially lived on subsistence.

Many older people in coastal Bangladesh suffered from various chronic diseases, although health was their most important asset as it affected their ability to work and maintain a minimum standard of living (HelpAge, 2002). They were unable to get healthy and nutritious food and adequate medications due to their financial hardships. They also depended on the local pharmacy to get medications in case of illness, as government health services in the area were inadequate. In addition, the health status of older people was consistently associated with their vulnerability, as it was generally associated with lower earning capacity (Dwyer et al., 2004). Further, their reduced physical strength and declining mental capacity were found to make them vulnerable to diseases (Malak et al., 2020). In the absence of proper basic initiatives - lack of social awareness, lack of education, limited access to health facilities - vulnerability borders on the extreme. Often, older people lack adequate sanitation and hygiene facilities, food security and family care and support, making them vulnerable to disease (Malak et al., 2020).

In sum, older people's vulnerability was the outcome of complex interactions of individual risks, namely being exposed to a threat, realising a threat and not having the defences or resources to deal with a threat. The vulnerability of older people was to some extent rooted in the geophysical location of the area, the housing structure and the location of their housing and livelihoods. Dependence on family members, especially sons and daughters, makes them more vulnerable in terms of basic needs. Poverty is one of the most important factors in the vulnerability of older people living in the region. It is often the result of cumulative disadvantages over a lifetime, as well as gender disparities and structural problems. Older people were often financially vulnerable, as their care and living costs could sometimes exceed their income and support. In addition, poor quality construction materials used in their houses, inadequate nutrition and a lack of social networks make them more vulnerable to extreme weather events.

References

Adger, W. N. (1999). Social vulnerability to climate change and extremes in coastal Vietnam. *World Development, 27*(2), 249–269.

Alam, E., & Collins, A. E. (2010). Cyclone disaster vulnerability and response experiences in coastal Bangladesh. *Disasters, 34*(4), 931–954.

Alauddin, M., & Rahman, A. (2013). *Climate change and sustainable development: The context of Bangladesh.* Bangladesh Institute of Development Studies (BIDS).

Andrew, M. K. (2010). Social vulnerability in old age. In H. M. Fillit, K. Rockwood, & K. Woodhouse (Eds.), *Brocklehurst's textbook of geriatric medicine* (7th ed., pp. 198–204). W.B. Saunders.

Bangladesh Bureau of Statistics (BBS). (2015). *Elderly population in Bangladesh: Current features and future perspectives.* BBS Statistics and Informatics Division (SID), Ministry of Planning, Government of the People's Republic of Bangladesh.

Barbosa, K. T. F., Oliveira, F. M. R. L. D., & Fernandes, M. D. G. M. (2019). Vulnerability of the elderly: A conceptual analysis. *Revista Brasileira de Enfermagem, 72*, 337–344.

Bourdieu, P. (2018). *The forms of capital* (pp. 78–92). Routledge.

Brocklehurst, H., & Laurenson, M. (2008). A concept analysis examining the vulnerability of older people. *British Journal of Nursing (Mark Allen Publishing), 17*(21), 1354–1357.

Cardona, O. D., Van Aalst, M. K., Birkmann, J., Fordham, M., McGregor, G., & Mechler, R. (2012). Determinants of risk: Exposure and vulnerability. In C. B. Field et al. (Eds.), *Managing the risks of extreme events and disasters to advance climate change adaptation* (pp. 33–78). Cambridge University Press.

Chau, P. H., Gusmano, M. K., Cheng, J. O., Cheung, S. H., & Woo, J. (2014). A social vulnerability index for the older people—Hong Kong and New York City as examples. *Journal of Urban Health: Bulletin of the New York Academy of Medicine, 91*(6), 1048–1064.

Crooks, D. (2009). *Development and testing of the elderly social vulnerability index (ESVI): A composite indicator to measure social vulnerability in the Jamaican elderly population.* [PhD dissertation]. Florida International University.

Dwyer, A., Zoppou, C., Nielsen, O., Day, S., & Roberts, S. (2004). *Quantifying social vulnerability: A methodology for identifying those at risk to natural hazards.* Geoscience Australia.

Eeuwijk, P. (2006). Old-age vulnerability, ill-health and care support in urban areas of Indonesia. *Ageing and Society., 26*, 61–80.

Fees, A., Varley, A., & Armas, I. (1999). After the earthquake: the potential for change in the built environment. *Environment and Urbanization, 11*(1), 121–136.

FRAC. (2019). *Hunger is a health issue for older adults: Food security, health, and the federal nutrition programmes*. Retrieved October 10, 2021, from https://frac.org/wp-content/uploads/hunger-is-a-health-issue-for-older-adults-1.pdf

Garschagen, M., & Schmidt, S. (2015). *World risk report 2015. Bündnis Entwicklung Hilft and UNU-EHS*. Retrieved October 10, 2021, from http://collections.unu.edu/view/UNU:3303#viewAttachments

Haque, C. (1997). *Hazards in a fickle environment: Bangladesh* (1st ed.). Springer.

Healy, J. D. (2003). Excess winter mortality in Europe: A cross country analysis identifying key risk factors. *Journal of Epidemiology and Community Health, 57*(10), 784–789.

HelpAge International. (2002). *State of the world's older people 2002*. Retrieved September 6, 2006, from http://www.helpage.org/Resources/Policyreports

HelpAge International. (2013). *Global age watch index 2013: Insight report*. HelpAge International.

Islam, M. A., & Rahman, M. F. (2017). Health problems of elderly people in Bangladesh. *Journal of Armed Forces Medical College, Bangladesh, 13*(1), 1–2.

Islam, M. R. (2018). Climate change, natural disasters and socio-economic livelihood vulnerabilities: Migration decision among the Char land people in Bangladesh. *Social Indicators Research, 136*(2), 575–593.

Islam, R., & Khan, S. M. (2018). Disaster vulnerability of the coastal poor in Bangladesh: an analysis of the livelihood and asset-based framework. *Disaster Prevention and Management, 27*(1), 74–87.

Islam, R., et al. (2018). A framework for building cyclone-resilient housing in coastal Bangladesh. *International Journal of Disaster Risk Reduction, 31*, 1–13.

Kabir, R. (2009). Understanding elderly vulnerability in Indonesia. *IIAS Newsletter, 32*, 12.

Khan, A. E., et al. (2014). Living in a floodplain: A study on the vulnerability of households to floods in Bangladesh. *Natural Hazards, 71*(3), 1735–1755.

Kreager, P. (2003). Understanding elderly vulnerability in Indonesia. *IIAS Newsletter, 32*, 12.

Malak, M. A., Sajib, A. M., Quader, M. A., & Anjum, H. (2020). "We are feeling older than our age": Vulnerability and adaptive strategies of aging people to cyclones in coastal Bangladesh. *International Journal of Disaster Risk Reduction, 48*, 101595.

Moser, C. (1998). The asset vulnerability framework: Reassessing urban poverty reduction strategies. *World Development, 26*(1), 1–19.

Paul, B. K., & Routray, J. K. (2011). Household-level recovery from floods in Bangladesh. *International Journal of Disaster Risk Science, 2*(1), 22–33.

Porter Starr, K. N., McDonald, S. R., & Bales, C. W. (2015). Nutritional vulnerability in older adults: A continuum of concerns. *Current Nutrition Reports, 4*(2), 176–184.

Reale, A., & Handmer, J. (2011). Land tenure, disasters and vulnerability. *Disasters, 35*(1), 160–182.

Salamon, S., & Lockhart, V. (1980). Land ownership and the position of elderly in farm families. *Human Organization, 39*(4), 324–331.

Sarvimäki, A., & Stenbock-Hult, B. (2016). The meaning of vulnerability to older persons. *Nursing Ethics, 23*(4), 372–383.

Schröder-Butterfill, E., & Marianti, R. (2006). A framework for understanding old-age vulnerabilities. *Ageing and Society, 26*(1), 9–35.

Shams, S., & Shohel, M. M. C. (2016). Food security and livelihood in coastal area under increased salinity and frequent tidal surge. *Environment and Urbanization ASIA, 7*(1), 22–37.

Thompson, J., Tod, A., Bissell, P., & Bond, M. (2017). Understanding food vulnerability and health literacy in older bereaved men: A qualitative study. *Health Expectations, 20*(6), 1342–1349.

UNISDR. (2004). *Living with risk: A global review of disaster reduction initiatives*. United Nations International Strategy for Disaster Reduction.

Weller, C. (2021). *Unemployed older workers face a tough job market*. Forbes. https://www.forbes.com/sites/christianweller/2021/05/12/unemployed-older-workers-face-a-tough-job-market/?sh=2e519c5c16e7
World Health Organization. (2018). Ageing and health [Fact sheet]. Retrieved June 2022, from https://www.who.int/news-room/fact-sheets/detail/ageing-and-health
Yoon, D. K. (2012). Assessment of social vulnerability to natural disasters: A comparative study. *s, 63*(2), 823–843.

Chapter 5
Livelihood, Land and Housing Vulnerabilities of Older People

> If there were a cyclone, I would have to leave everything behind and move to the cyclone shelter to save my life. Everyone had to leave their house. When I returned from the shelter after a cyclone, I found nothing. I had to start from scratch again.
>
> - An eighty-year-old participant from Padma village.

5.1 Introduction

This chapter analyses how extreme weather events have increased the vulnerability of older people in the study villages. The coastal region of Bangladesh is vulnerable to extreme weather events - cyclones, floods, tidal waves and soil erosion. The coastal population of Bangladesh also faces various forms of extreme weather events simultaneously, which cause heightened vulnerability (Paul & Rashid, 2016). The major extreme weather events in the study area include strong cyclones, floods, erosion of riverbanks, hot summers, bitter winters and tidal surges. Some elderly residents reported that bitter winter weather was worse now than in the past. Others observed that summer was getting hotter by the day. Extreme weather events have a negative impact on the vulnerability of older people in the coastal areas of Bangladesh. Strong winds and floods have destroyed their houses, roads, bridges and other facilities. Electricity and communication links were disrupted, and highways and rivers were impassable. Drinking water was contaminated by waste and salty water from the floods. In 2007, for example, cyclone SIDR caused significant damage to infrastructure and assets, as well as lost production worth BDT 30.2 billion (US$ 438 million), 95% of which occurred in the affected coastal areas of Bangladesh (GoB, 2008).

Older people face complications, suffering and damage after extreme weather events, especially after a cyclone. They were often unable to cope with the aftermath, as there appeared to be insufficient resources and relief services. Concerned government officials, local and national humanitarian organisations and NGOs helped and

M. I. Ali et al., *Extreme Weather Events and Vulnerability of Older People*,
https://doi.org/10.1007/978-3-032-11256-9_5

provided support in the aftermath of the cyclone, but the capacity, resources and steps required for emergency relief were unevenly distributed. Political and social rank and power imbalance also played a crucial role in exacerbating the vulnerability of older people following various extreme weather events (Mallick et al., 2017). This chapter analyses how extreme weather events increased older people's vulnerability by affecting their employment conditions, income and expenditure, housing and living conditions, land and assets.

5.1.1 Livelihood Vulnerability

Extreme weather events negatively affect the livelihoods of older people in coastal regions, making them financially vulnerable. Most older people did not have formal or regular jobs. Agriculture and fishing were the main sources of employment, followed by some small businesses in the region, which had consistently been affected by almost every extreme weather event in the past. Household members of older people were engaged in fishing in the sea and coastal rivers and collecting wood and honey from the mangrove forest - Sundarbans. Older people who could engage in these activities with their sons, brothers or nephews could earn some income. Cyclone SIDR in 2007 caused significant damage to about 2.3 million households, of which about one million lost their livelihoods. As many households had overlapping livelihood activities, they were simultaneously affected by losses in crops, livestock, fisheries, trade, industry and employment (GoB, 2008).

5.1.2 Loss of Employment and Income

Almost all older people in the region reported that extreme weather events severely damaged their employment opportunities. Most had no or limited opportunity to engage in paid work, yet they had to do something to survive. Being a coastal region, most available jobs were related to fishing or irregular casual work. As they did not have regular jobs, they suffered immediately after extreme weather events. Most were unemployed immediately after the cyclone because the hazards were too great after the cyclone. Chandresh, a 77-year-old widower from Haritana village, explained:

> After cyclones and floods, we have to remain unemployed. There is water everywhere, and trees have fallen on the roads here and there. Dead bodies of people, animals and birds – cows, goats, poultry – were found everywhere [in cyclone SIDR]. It was difficult to move around and find work.

After each extreme weather event, older people in the area became either unemployed or underemployed. After the events, it was difficult to move from one place to another and communicate with people and potential employers. This was because extreme cyclones destroyed all infrastructure – roads, dams, homes and communication systems. Trees were uprooted on the roads and streets, hindering people's daily movement to or from work. The tidal surge and cyclones inundated the entire area, causing days of flooding and forcing elderly people to stay at home without work. Chandra, a 73-year-old farmer and casual labourer from Charlathimara village, said:

> I lost my work after the extreme events – cyclones – in the past. During the Cyclone SIDR in 2007, I could not do any work for 8 to 10 days. All [roads, paddy fields, farms and fishing infrastructure] were underwater [due to storm surge and flooding].

An 80-year-old Rashidul from Haritana village said he too was jobless. "In the last cyclone in 2019, the water in my courtyard was up to my waist and it took two to three weeks to dry. How could I work under those conditions?" he said. For most, it was not possible to work. The same happened during Cyclone SIDR and other cyclones in the past. After the cyclones and floods, they were unemployed due to limited employment opportunities. Those who were self-employed had to stop working for days because of the destruction of workplaces:

> I could not do any work at the time of [the] cyclones. I had no work at that time. At the time of Cyclone SIDR, work was stopped for many days. Probably for 2 months (a 62-year-old man, Biddut, a casual labourer from Padma village).

Work opportunities also declined sharply immediately after the cyclone, leaving them without work for weeks and even months. Elderly people who sometimes worked as regular casual labourers lost their employment after the extreme events, as their workplaces collapsed. Some participants said that the bitterly cold weather after the extreme weather events prevented older people from working because they were physically unable to cope with the adverse temperatures. As older people did not have enough strength to beat the harsh cold wind after the events, some employers did not want to hire them for work., Rashidul from Haritana village claimed, "… no one gave me work in the cool weather [after the extreme event]. They said this old man could not work in this bitter cold".

Community life also came to a standstill after any extreme weather event, especially after a cyclone. The waterlogging after the cyclone prevented agricultural activities for months and sometimes even years, as the soil was contaminated by the inundation of seawater. Many participants were out of work in labour-intensive sectors such as agriculture and fishing in the region due to the damage. The loss of boats and equipment further limited the fishing capacity of those who were engaged in fishing, which dramatically affected the job prospects and earnings of older labourers. Shamiul, an 80-year-old man from West Haritana village, described:

> At that time [after the cyclones and floods], everything [work and activities] almost came to a standstill. I did not go anywhere. Where was I supposed to go? All the places were flooded. Rivers and canals overflowed their banks. I could not work then. This always happened when there were cyclones and floods.

Shamiul had a small fishing business. He had a fishing boat and nets that were washed away; after Cyclone SIDR in 2007, there was nothing left for him to work with. Older people who were marginal farmers in some way lost their farm implements – ploughs and animals, which also prevented them from going to the fields, reducing work opportunities. However, the duration of unemployment depended on the strength and severity of the cyclone and floods. Sabdar, from Padma village, said:

> [...] Work remains on hold during cyclones and floods, but it depends on the magnitude of the cyclone and the extent of the floods. If the weather events are less sustained, work will stop for a short time. On the other hand, if the cyclone was severe and the floods were of longer duration, work stopped for a long time.

Increased unemployment after cyclones and floods dramatically reduced the income and wages of older people and their households. Since most older people had limited capacity to work, they had no or inadequate income to survive. A 77-year-old participant, Chandresh, said:

> I had no income [after the cyclone]. [...] my sons could earn some money if they had work. He used to cut trees on the roadsides as a casual labourer. After one or two days of work, he remained unemployed most of the time because there was no work [which was the only source of income for the household].

The older people reported that they had lost their regular daily income due to the unavailability of work after serious events in the past. They could only earn some money when they were able to work. Chandra, a 73-year-old man from Charlathimara village, said: "There was no regular income [during the last cyclone]. There was no certain income per day. If I could do some work, I would get some money". Depending on the availability of work and their capacity to do the work, their income ranged from BDT 10 to BDT 500 per day. However, they could not receive wages for some time after the disaster. The drastically decreased employment opportunities following an extreme event affected their income. For most, their post-disaster income was not enough to meet their basic daily needs, which increased their suffering for days. Older women who were divorced, widowed or separated had even fewer income-generating opportunities after an extreme weather event and were one of the hardest-hit groups in the community.

5.1.3 Increase in Family Expenses

Due to the decline in employment and income, the extreme weather events affected the expenditure of older people and their households in the region. Maintaining household expenditure was a critical issue for older people in coastal areas after the disaster. Although their expenses were limited, it was difficult to manage their inadequate income and support. Most reported that their expenses exceeded their income after extreme weather events in the past. They could not buy the daily necessities. Rashidul from West Haritana village said, "Daily expenses were higher than my income. With BDT 100, I could buy five kilograms of rice. I was struggling to manage my [other] daily expenses". Chandresh from Haritana village said:

> My expenses are very low. I drink some tea every day. However, my household expenditure is BDT 150 to 200 per day, with an average of BDT 4000 to 5000 per month. Income is not enough to cover household expenses. Sometimes it is difficult to arrange oil for cooking. Again, it is very difficult to manage household expenses during a time of disaster. If we do not get help from GO/NGO, the expenses become too great to manage.

For some, following any extreme event, the cost of living gradually increased due to higher prices. Shamiul, a 72-year-old fisherman from West Haritana village, said that expenses had increased significantly during the disaster. Older people who lived in extended families said their income was not enough to meet daily expenses. Chandra, a 73-year-old from Charlathimara village, claimed:

> The expenses were increasing every day. There were ten of us in the family, including grandchildren. Expenditure was very high at the time of the disaster as basic commodities became more expensive.

The older people of the area struggled to make a living in the past when they were affected by extreme weather events. Most of their income went towards repairing their houses and buying food and clothes after an extreme weather event. They also had to pay for health costs, including the cost of medicines and medical care, and sometimes the costs associated with the funeral after the death of family members in the cyclones. Most older people said that they would not have been able to meet their living expenses after the extreme events if they had not received assistance. Sometimes they could not buy food and cooking utensils if they could not find work. They said that it was very difficult for them to manage household expenses at the time of the disaster. Elderly participant, Chandresh, a 77-year-old widower from Haritana village, said, "... managing household expenses during times of extreme weather events was challenging." A 67-year-old Hossain, a marginal farmer and casual labourer from Badurtala village, said, "The daily expenses increased after the disaster because everything was washed away; [almost] nothing could be bought [near us]. Only the brick-built houses survived the cyclones and stood here."

There were several reasons why the daily expenses of older people and their households increased. During Cyclone SIDR, almost everything was destroyed in the area, including houses, shops, and roads. Most areas were flooded, trees were uprooted on the roads and streets and the supply of goods was disrupted for days. The expenses increased after the cyclones and floods, and they sometimes had to take loans to cover their daily expenses.

5.1.4 Increase in Debt

Another financial vulnerability of older people and their households after an extreme weather event was debt. As older people and their households were forced to borrow to meet their immediate needs and livelihoods due to the cyclone, they were heavily indebted. Many households took loans from landlords or from other people to replace lost livelihood assets such as chickens and goods to restock shops (GoB, 2008). It was noted that older people and their families found it difficult to cope with increased household expenses with the limited income they had and the support they received, which pushed them into borrowing. Most had to take loans from various banks, NGOs and moneylenders. Fakirshah from Padma village said that he took loans from NGOs, fishmongers, moneylenders and banks. He reported that he borrowed nearly BDT 0.4 million from moneylenders and BDT 30 thousand from the Agriculture Bank and the rest of the loans were taken from different NGOs – ASA, Sankolpo and Sangram. He also said that he was facing difficulties in repaying the loan instalments. The interest on his loan was still increasing as he had not been able to repay the loan. Some took a loan to start their business from scratch after the cyclone, and some others reported taking loans to repair their houses. Fakirshah said again:

> I took the loan before and after Cyclone SIDR. I spent it on buying nets and trollers [motor-driven fishing boats] for the fishing business. I did not have a big business anymore. I ran a small business by taking a loan.

Some took loans to build a new house or repair their destroyed house after the disaster. An older participant, Shamiul, a fisherman from West Haritana village, described:

> I took a loan after the Cyclone SIDR and another loan from the Agricultural Bank before the SIDR. I repaired my house with that loan. I bought the tins (corrugated sheets). Sangram NGO built a house for me after the SIDR. They gave me 20 pieces of tin after the SIDR. I bore all the other expenses myself.

Some elderly people took a loan to start a new business or rearrange their business activities and for medical expenses, while others borrowed money for food and daily needs. Some older people took loans for fish farming in their ponds. One participant, 66-year-old fisherman Sabdar from Padma village, said:

> I took a loan of BDT 0.13 million to raise fish in my pond. After a cyclone, I again took BDT 10,000 from an NGO. I have not been able to repay the loan yet. The interest on this loan has increased to double the principal amount now and continues.

However, he reported that all the fish in his pond had been washed away during the SIDR cyclone. There was nothing left to repay the loan. Most of them did not pay the dues on time due to their financial incapacity and increased unemployment, or their inability to engage in income-generating activities. Some older participants were in debt and had been sued for repayment. One mentioned that he spent nearly BDT 40,000 on the construction of his house, including corridors on all sides. He had taken the money as a loan from Sangkalpo and Songram (NGOs) but could not repay it. He said that a court case is pending against him for not repaying the loan. Similar comments were made by Chandresh, from Haritan village:

> … Yes, I had a loan from Polly Schanchy Bank (Village Savings Bank) – [a government project]. I got BDT 7000 but could not repay the dues. I had repaid only two dues. I had no income to repay the loan. If I do not pay off the loan, they might send me to prison. I will go to prison, as I can't repay the loan. […] no bank or NGO would approve a loan for me because I am unable to repay it. I had bought some tin and the necessary materials to repair my house; some money I spent on healthy food and clothes.

Generally, older people had limited access to the credit market due to their limited income, assets, employment and physical condition. Most older people reported that they were unable to repay the loans they had taken at different times – before or after the cyclones – for different purposes and from different sources. A 60-year-old widow, Raima, from Western Badurtala Cluster Village said, “I had borrowed some money. I took 40 thousand BDT from the NGO Sangram. With the loan, I built this house and bought this tin [material to repair the house].

She also owed nearly BDT 7000 to two shopkeepers in the local market. She said, “I am in debt of about BDT 10,000 to shops and other people.” Borrowing from various sources was seen as one of the coping mechanisms after the extreme events to address the losses and damages, although the government, national NGOs and foreign donor agencies provided some assistance throughout the recovery process. However, the recovery efforts were not enough for most of the older people to recover from their losses. As a result, the poor older people and their households had to borrow money from formal and informal institutions such as microcredit providers, banks, moneylenders or other sources (Subhani & Ahmad, 2019). But when the time came for repayment, there were problems for the elderly. Many were not able to repay the borrowed money. Moreover, most of the older people and their families reported that when they borrow money from private businessmen (Aratdar), they have to sell their fish catch back to them below the market price. They reported that they were under financial pressure from the moneylenders to continue repaying their loans. They often ignored cyclone warnings and stayed at sea in extreme weather conditions, keeping them vulnerable to death (Government of Bangladesh, 2008).

5.2 Loss of Land and Assets

Extreme weather events destroy the assets of older people and their households almost every year. The damages following a catastrophic event included the loss of fishing boats and equipment, production facilities and tools for self-employed people. Some older people said that the cyclones and floods destroyed their fishing ponds and fishing gheras (enclaves or fencing) and that they lost their family’s livestock and birds. Public infrastructure such as roads, dams and power grids were also destroyed. The cyclone and floods destroyed their cowsheds and other agricultural infrastructure. Siltation of land, soil erosion, loss of inventory, destruction of shops or business premises and loss of stored food were other consequences of each extreme weather event in the region. According to the Joint Damage, Loss and Needs Assessment (JDNLA), the total damage and losses caused by cyclone SIDR in 2007 were estimated at (US$ 1.7 billion) BDT 115.6 billion (GoB, 2008; Islam, 2018).

5.2.1 *Loss of Land*

Patharghata – the coastal area where this study was conducted– is surrounded by rivers, along the banks of which older people and their households face various extreme weather events every year. Due to the negative impact of these extreme events, they lost their farmland and residential land. Most older people had no land – neither farmland nor residential land. Some of them owned small pieces of farmland or residential plots. Some were living on the state's khas land (the state's land) or the embankment. Others in these poor areas had inherited some land from their parents or grandparents, but most said the land was lost or damaged due to soil erosion caused by storm surges and floods, forcing them to become landless. A 60-year-old widow, Shakima, from Padma village said in tears: "I did not have a single piece of land. After my husband died in 1988, I lived with my father. My father had two acres of land, which we lost due to the erosion of the riverbank".

The destruction of the riverbank was one of the most drastic consequences of the cyclone's wave surge and flooding in the region. The socio-economic impact of this calamity on the elderly and their households was extreme. Three rivers surround the study area – Bishkhali, Boleshwar and Horinghata. Some of the older people and their families have been affected by the riverbank erosion of these rivers several times in the past. The erosion of the riverbanks has mainly affected the livelihood of the farmers, as both their agricultural and residential lands were lost. Amena, an 80-year-old widow from Padma village, reported:

> My father was a rich man. After my marriage, I found that my father-in-law was also very wealthy. My husband owned five acres of land, all of which has been lost due to erosion of the riverbanks. Most of my father's land was also lost to erosion.

The 60-year-old widow Palkey lives with her son in Ruhita village near the dam, but the house is not hers. She lost her house in Cyclone SIDR in 2007. She recounted her horrific experience, "My house was on the bank of the river near the dam. It was destroyed by the breaking of the riverbank at the time of the SIDR in 2007. I now live on other's land." Some older people said they had to move to nearby roadsides, embankments and surrounding towns when they lost their land and residence. Rajina, a 63-year-old tea seller, lives with her husband in the same village of Palkey on the embankment. She said:

> I do not have land anymore. The land I had was lost to the Ganga [river] due to erosion of the river. I had ten kathas of land in the field. The land yielded some paddy. I had a house. There were many coconut trees in my house. All of them were destroyed by the collapse of the riverbank.

Lailunnahar, a sixty-year-old woman from Padma village, said she had no land as she had not received anything from her father or husband. She said she had just got two kathas (nearly 6 decimals) of land from her mother where she lives now. "My husband had only 60 decimal lands. But it was eroded in the river five/six years ago," she added. Eighty-year-old Abdullah of Charlathimara village said his father used to cultivate some land. But the river had grasped all their land in the past. He claimed:

> The river has also grabbed our home. I do not have a house anymore. I live in a government-provided house. The Bangladesh Navy built it. But that will also collapse soon. The place where I live is not safe and secure. It could fall into the river at any time.

Some of the respondents said they had to sell their land to meet their family expenses in the aftermath of the cyclone. Enam, an 80-year-old widower from Padma village, said he was a farmer and made a living from cultivating his land. But he also lost his land. Now he is landless. He stated:

> [...] But after the disasters, there was an acute food shortage, there was no food left at home. What were we supposed to eat? The food rations provided by the government were taken only by the powerful people. We could not manage with what they distributed to us. It was only enough for two days or one day [what we get]. Then we had to meet the expenses ourselves, otherwise, we had to sell our land properties.

Losing their farmland and residential land puts them at risk of becoming homeless. They then have to move their house to government-owned land or to their neighbour's land temporarily, which greatly affects their socio-economic life and livelihood. Sometimes they or their households have to migrate to a nearby town or big city losses that make the older people vulnerable to living alone.

5.2.2 *Loss of Resources*

Although farming and fishing were the main sources of income, livestock rearing was also an important source of income for the older people and their households in the region. They generated daily or monthly income from the sale of milk, eggs, poultry and goats. However, the extreme storm surge from the cyclones drowned their animals and birds and destroyed the livestock sheds or uprooted trees, causing them to lose their livelihoods. Biddut, from Padma village, said:

> I had 100 feet of the fishing net, but everything was damaged, and my fishing boat was destroyed. Did I not tell you that I had only one piece of cloth – which I was wearing – left? ... I had cows and goats; I lost them all.

The elderly fishermen and their household members had been affected by the extreme cyclone several times in the past. Some said that the cyclones also destroyed their fishing infrastructures - ponds, fishing gheras (enclave or fencing) and equipment. Their fishing boats and trollers (motorised fishing boats) and fishing nets were destroyed or washed away; the fish from their ponds were washed away by the overflow of water from the storm surge. An elderly participant, Sabdar – a small farmer – from Padma village reported:

> The fish in my pond (worth about BDT 0.1 million), a shop and a troller were washed away during cyclone SIDR. I had six bags of fishing nets to catch Hilsa fish and many other items – all were gone, and nothing was found after the cyclone.

During the cyclone, every one of the fishermen lost their fishing boats, nets, fishing gear, the undercarriage of the boats and other materials in fishing boats. As a result, they could not go out to sea to fish after the cyclones. A fisherman from Padma village said he had been involved in fishing since his youth and lost all his fishing equipment due to the SIDR cyclone. An older person also said that he had bought all his fishing equipment with loans from various sources. He was upset when he talked about his loss at the interview. Most of the older people said that during the cyclone, various animals and birds such as goats, cows, chickens and ducks were washed away or died. About 80% of animals and 76% of birds died in the four worst-affected districts in the coastal regions during cyclone SIDR in 2007. Animal and poultry farms, livestock, food, fodder and grazing land were also lost during this cyclone (GoB, 2008). It also destroyed crops in the fields of some older people. A 67-year-old marginal farmer, Hossain, from Badurtala village described:

> … In the SIDR, there was nothing but us – the members of our family. During cyclone SIDR, it was found that everything – houses, farms, paddies, livestock and poultry – had been blown away and washed away. Everything in my house had been destroyed.

The older people who were small retailers in the affected areas had lost almost everything. The small factories (rice mills, sawmills, potteries), shops (blacksmith shops, barber shops, repair shops of various kinds), some wholesale shops, several repair shops for tricycles and rickshaws, sewing machines in tailoring shops, sewing machines in private houses and vegetable shops were destroyed. Retailers in the areas affected by the recent extreme events witnessed the destruction of their business premises and the loss of inventory in their shops. A small shopkeeper, Kamal, from Badurtala village, said:

> I lost all my inventory; all the goods in the shop were destroyed and washed away, and my shop collapsed in Cyclone SIDR. My shop was right next to my house. All [shops and houses] collapsed. I had no place to live. […] I was selling clay pots and groceries, but all of them were washed away with other goods. My oil drums and containers were also washed away.

Older people in coastal areas have had bitter experiences of cyclones and floods. Some participants said that almost everyone lost their belongings in the last extreme cyclone. Older people shared their stories of suffering that were due to the loss of various material goods such as food, bedding, toilets and other possessions of older people and their households. Those older people who had some cultivable land had lost their crops. An older woman participant, Shakima, a widow from Padma village, reported: "[…] During cyclone SIDR, I was with my daughter. All my clothes, fishing nets of my son-in-law – everything – even the pot I kept rice in was washed away".

In addition to the losses from the severe cyclones and floods, the older people often had to sell their belongings to survive after the disaster. They claimed that if there was anything left after the cyclone, sometimes, they had to sell their bare necessities like land, livestock, building materials and other personal belongings to cope in the aftermath of the extreme weather events, making them even more vulnerable.

5.3 Loss of Dwellings

The vast majority of coastal residents in Bangladesh were living in impoverished and poorly constructed houses. Older people in the study area lived in houses made of corrugated iron sheets (tin) and wooden or bamboo posts and straws. Participants said their houses were inadequate and did not protect them from the strong winds of the cyclone and storm surge. Although most of the houses in the region were donated by the government and NGOs after the severe cyclone SIDR in 2007, some older people said they did not receive any support from the government or NGOs to rebuild their houses after the extreme weather events. Rather, they rebuilt and repaired their houses on their own and with the help of loans. The older people complained that the roofs of their houses, donated by the government and NGOs, were defective and leaking due to rust. In the rainy seasons, rainwater could enter their houses through holes in the roof. Some older people and their families lived outside the embankment and thus they became the first victims of the extreme weather events.

5.3.1 Loss of Houses

Cyclones caused damage to houses in three ways: storm surge, flooding and wind. Brick-built houses were much more wind-resistant and often served very well as local ad hoc shelters, but were still destroyed by flooding and storm surges. Storm surge was by far the strongest destructive factor in some areas, leading to the destruction of houses and non-engineered structures. The wind caused some direct destruction, which was generally quickly repaired, as well as a significant number of fallen trees – one of the main causes of the destruction of houses by the wind (GoB, 2008). Most older people shared their experiences of losing their houses after the cyclone. Some of them said that their houses were destroyed; some houses had collapsed, and others were blown away or washed away. One participant, Fakirshah, reported:

> Yes, my house was destroyed. There was a big loss during the SIDR. I could not identify the place where my house was. The items in the house were also washed away. The foundation of my house collapsed. It also collapsed during the last Cyclone Bulbul in 2019. I have not [re] built it yet. Bulbul blew off the roof of my house. The trees nearby fell and crushed [my] toilet.

The older people of the area believed that extreme weather events such as cyclones, storms and floods have increased over time. Some said that they had experienced more than one cyclone and flood in their lifetime and that their houses had been destroyed more than once by past cyclones. Rashidul, a marginal farmer from West Haritana village, described his situation:

> … My house has been destroyed six times by cyclones and floods at different times [in the past], most recently by Bulbul in 2019, when trees fell on my house. I have not been able to rebuild it yet, no [other] way to put beds and bedding [in the rooms of my house].

Similar statements were made by some older people from other villages. Most participants reported that their houses had been hit by cyclones three or four times. Since most of the older people and their families were not able to build a house with bricks, they tended to use wood and thatch. Many reported that only the foundation walls of their houses were left. For others, the houses still stood after the cyclones but were not habitable because of the damp and muddy floors. Older people used dry materials when possible to make them usable for themselves. An 80-year-old Enam, a widower from Padma village, described:

> There was a major problem after floods and cyclones. The house floor becomes muddy after the houses are flooded. There was a sleeping problem. So, the foundation of the house was flooded in the last cyclone. Then we spread ashes on the damp ground, especially ashes of cinders. We had to spend several nights in this condition until we could repair it.

The extremely precarious situation older people found themselves in after the cyclones often meant they had nowhere to live. Some reported that they had to stay out in the open under the trees for days until they found their accommodation or received help from aid organisations. Many told how they had temporarily repaired their shelter with uprooted trees and plastic sheets after the cyclone and stayed there until they managed to build a permanent frame. One 70-year-old participant, Abul, from Padma village, explained:

> My house was swept away during Cyclone SIDR in 2007. After the cyclone, I stayed here and built a polythene shed. After three months, I got some corrugated sheets. After a few months, my house was [re] built with this sheet [corrugated metal] and wooden posts.

Rizeeya, a 70-year-old widow from Purbo Badurtala village, experienced similar conditions. She explained:

> The house is made of wood and corrugated iron. The sheet metal has been badly damaged by cyclones and floods. I am living in a makeshift polythene shed. I still cannot repair the door and put it up. I have no way to repair it. That is why I cannot do it.

Older people from the affected villages reported that it was very difficult for them to rebuild their houses after the cyclone. Some received material and non-material help from various sources. Others, however, claimed that they did not receive any assistance at all after the cyclone. A 69-year-old respondent, Raizuddin, from Ruhita village said:

> Cyclone SIDR caused a lot of damage. My house had collapsed. There was no shed on top of the roof anymore. It had been blown away. Later, I rebuilt and repaired it with wood and corrugated iron. I had not received any support. It cost me BDT 20,000. I used my savings to cover this.

Some elderly people reported that some parts of their house – veranda, roof, or wall – were damaged in the past cyclones. Some did not receive the necessary assistance due to problems with the management and implementation of relief programmes, including the programmes' association with certain political and social ideologies. Elderly people who did not receive assistance from the government and non-governmental organisations had to borrow money from the bank, NGOs and moneylenders. A 60-year-old widow, Raima from Western Badurtala Cluster Village, pointed to a nearby place during an interview and said:

> … Here was [my] house, which was destroyed [by the cyclone]. I had no place to live. I did not get a house as compensation. Maybe it had been two years since I had repaired my house. Before that, I had lived in a broken house for 11 years [repairing the house]. It cost nearly BDT 0.2 million to repair and build that house. My son borrowed money from a bank and got some [financial] help from his father-in-law.

It was estimated that 34% of households with destroyed houses and 17% with partially destroyed houses live in the area affected by Cyclone SIDR (GoB, 2008). The destruction of houses and loss of housing affect the living conditions of older people in the region.

5.3.2 Suffering on the Way to the Cyclone Shelter

Reaching and staying in shelters was one of the vulnerabilities faced by older people during extreme weather events in the past. Older people in the region said their houses were not strong enough to protect them from the devastating cyclone. Therefore, they had to find a safe place to shelter and usually went to the cyclone shelter, although the number of cyclone shelters was not enough to protect them all. But they faced many dangers on their way to the shelters. Older women, especially older widows, were more vulnerable to self-evacuation than their peers. One older widow recounted how she and her daughters were evacuated on their way to the shelter during a past cyclone. Raima, a sixty-year-old widow from West Badurtala village, narrated:

> My house was no longer safe. I did not dare to stay at home when I heard the news of the cyclone. […] I climbed a tree near my house with my three daughters and tied them up. But I was not able to protect all of them from the strong onslaught of the water. I left my disabled daughter in the water. Although it was very sad to lose a daughter, there were no alternatives for me. [With tears in her eyes, she said again,] I thought one of them would let go. After a while, the daughter came back to me swimming and said: Ma [mother], I have come back.

But when she came home after the cyclone, she found that they had nothing left. This kind of suffering was common for older people and their households during extreme weather events. The government's disaster management office, with the help of members of the Bangladesh International Red Crescent Society and local volunteers, made efforts to announce the signal strength and severity of the cyclone over loudspeakers. They said most people did not want to go to the shelter until the cyclone hit the community. In their collective experience, women were more vulnerable to self-evacuation. The 67-year-old Kariman from Western Badurtala Clustered Village described:

> During Cyclone SIDR, there was extreme destruction. My mother and I were in my house at the beginning of the cyclone. But when it was no longer possible to stay here, we wanted to go to the shelter. On the way to the shelter, the storm surge emerged suddenly. The roar of the water made us fall into a pond. There was a coconut tree on the bank of the pond. We

> held on to the coconut tree in distress. […] I cannot remember how I climbed up and down and reached the centre. When I returned from the shelter the next day, I saw nothing but the destroyed foundation of my house.

Older people staying in the remote area of the village were particularly vulnerable. Since most of them lived in high-risk areas, there were hardly any good roads and communication facilities. A 60-year-old widow, Palkey, from Ruhita village, recounted her horrific experience:

> As I thought we could get into trouble, I was holding my elderly mother's hand and taking my baby on my lap during the cyclone. I came by holding my mother's hand. At that time, a cotton tree was uprooted over the road. My mother went out of my hand. It was quite dark, and I could not locate my mother. Then I returned to the embankment with my children and took refuge in a shop. After that, I noticed a man brought my mother from the other side of the embankment half an hour later. If he had not helped, my mother might be dead. We spent the whole night in the shop. In the morning, after the cyclone passed, I saw the foundation of my house level with the ground of the flat land.

Most of the time, older people have to go to the cyclone shelter alone or with the help of others. Sometimes older people get injured by the cyclone during the evacuation. A female participant, Shakima from Padma village, described her situation during the cyclone: "During cyclone SIDR, my house collapsed on the road next to Tofazzal's (a neighbour) house. I was injured by a falling pole when I went to the shelter. It still hurts me." It was extremely difficult for older people in the area to go to the shelter or any other safe place during the cyclone. Many of the older people said that the water deluge and wind prevented them from going to another safe place. The surfaces outside their homes also made it difficult to go to another place. Walking around the village, it was noticeable that the pavements, paths and nearby surfaces were far from ideal for the mobility of older people. There were also ponds in front of their houses, which made things even more difficult. There were several houses in remote areas that could not be walked through. In addition to the environmental hazards, limited mobility could cause significant problems for older people during cyclones and floods. If they were left behind or unable to reach vital facilities, they faced enormous difficulties.

Although the cyclone shelters were the safest place during the cyclone, their number and accommodation capacity were not sufficient. Most of the cyclone shelters were located within half a kilometre to three kilometres from the participants' houses, but the roads and communication routes were not easily accessible for the elderly. The conditions in many of the shelters did not meet the needs and requirements of older people. An elderly male participant, Chandresh, from Haritana village, provided the following statement, which reflects the views of many older people:

> The living conditions in the cyclone shelter were not good. All the people – children, youth, women, men and elderly – had to stay in one place. We go there as soon as possible when a cyclone comes, with one or two sets of clothes. There were no special arrangements for the elderly. There were two toilets, one for men and one for women … this was not enough for the number of people in the shelter. There were no provisions for drinking water. We had to fetch drinking water from outside the shelter, but the quality of the water was not good. The condition of the building was also not safe enough. In some areas, it was no longer safe to stay there. It was also not possible to have privacy there.

Most believed that they would be protected from severe cyclones and floods if they went to the shelters. However, others felt that some cyclone shelters were risky, as most were outdated and defective. Again, most shelters were stretched beyond their capacity during the cyclone. A senior participant, Fakirshah, said:

> I used to go to the cyclone shelter during cyclones Bulbul and SIDR. But now I think it is risky. It [the shelter] could collapse if it is loaded with too many people. It was unnecessary. There were many more people here than it could accommodate.

Due to the poor condition of the public cyclone shelters, inadequate warnings and individual perceptions and beliefs, some elderly people and their households refused to evacuate, believing that their houses could withstand a cyclone. Some elderly people went to well-built homes nearby or took shelter during the cyclone in the nearby schools, shops, on the embankment or under a huge tree. For example, the Fakirshah said, "After the cyclone shelter was filled, we had to stand in the shops and [sometimes] on the embankment." An elderly person from West Haritana village added: "I went to my brother-in-law's brick house and was there every time during the cyclone". Another elderly person from Charlathimara village said that he did not go to the shelter but if necessary, he went under a strong tree and held on until the cyclone passed.

The cyclone shelters were open to all affected people at the time of cyclones and floods. Some older people stated that there was a lack of space in the shelter and privacy. There were no separate rooms with beds and bedding, food, water and mobility for the elderly. There were also no medical facilities and high noise levels. The shelters were overcrowded with young and old men and women. However, volunteers often helped. An elderly person explained:

> [...] The volunteers first took the elderly people in the shelter and put them in a corner, but the food was the same for everyone. Everyone got help and everything they distributed. Drinking water was not available there. There were two toilets. A tube well [with water] was also available. But the water was not very healthy. The conditions in the cyclone shelters were not adequate to allow older people to stay comfortably. Some older people felt that the cyclone shelter was not built with the demands and needs of older people in mind, but was rather seen as a socio-political asset of the community.

The experiences of the participants indicate the various losses and vulnerability of older people and their households to extreme weather events in the coastal areas.

5.4 Discussion and Conclusion

Bangladesh experiences various extreme weather events such as cyclones, floods, extreme temperatures, flood and storm surges, tornadoes, thunderstorms, riverbank erosion, salinisation and coastal erosion every year (Burkart & Kinney, 2017; Masum, 2019; Paul, 2015). The experiences of older people in the coastal region of Bangladesh show that extreme weather events increased the vulnerability of older people by affecting their employment and working conditions, their income and expenditure, their ability to borrow, and their homes and assets. Similar results have been found in other studies on the impacts of climate change and extreme weather events on humans and the environment in the region (Saha, 2015; Islam & Walkerden, 2014).

Extreme weather events, directly and indirectly, affect the employment and income, credit liabilities and expenditure of older people in coastal regions. As a result, this group is particularly vulnerable. First, the availability of work was found to decrease sharply after the extreme events due to the destruction of sources of employment such as agriculture and fishing. Older people and their household members had little or no work after the cyclones and floods. Second, access to work was reduced by various environmental and structural hazards such as damage to bridges, roads, houses, factories and communication networks, as well as loss of equipment for self-employed people. A study on 'Disaster Response and Risk Management in the Fisheries Sector' (Westlund, 2007) found that natural disasters that occur suddenly and without warning can lead to significant losses and damage. These include the loss of boats, equipment, fish cages, aquaculture, farmed stocks and other productive assets, the destruction of working infrastructure such as landings and fish processing facilities, and production losses such as the escape of fish from fishponds (Westlund et al., 2007, p. 18). World Bank research shows that the average industrial firm ceased operations after the cyclone for 41 days, resulting in a loss of $3.65 million and 1.2 million working days.

This study showed that increased unemployment significantly reduced the incomes of older people and their families. Post-cyclone environmental hazards limited their access to work, which affected their income. In addition, their lack of physical strength and capacity meant that they were sometimes underpaid in the aftermath conditions. These factors contributed to the financial vulnerability of older people in coastal Bangladesh. Previous studies have shown that extreme weather events reduced agricultural productivity and employment, so households needed external assistance to cope with the loss of income (Mallick, 2014; Subhani & Ahmad, 2019). Cyclone AILA in 2009 directly threatened the income-generating capacity of coastal communities for several years, and many people from affected households went in search of income-generating opportunities to alleviate their hardship. This economic vulnerability was more severe among older women and widows compared to men. For example, one study found that poor older women were more vulnerable to inadequate care and support because they had limited financial and material resources and lived in generally impoverished environments (Eeuwijk, 2006).

The extreme weather events had an adverse impact on the family expenditure of older people in the region as it affected their jobs, income and production. Older people in the coastal region endured financial difficulties in managing their daily expenses immediately after severe weather events. The expenditure of the households of older people exceeded their income. Cyclone-triggered production, wages and employment were insufficient to cope with simultaneous commodity price and expenditure trends (Jahan & Haque, 2016). As the elderly and their households had no savings and reserves, they were unable to meet their basic daily needs without financial and material support from other sources. In the coastal region, prices increased steadily during cyclones and floods compared to the non-cyclone period. There are several reasons why the daily expenses of older people and their families have increased. In the context of vulnerability and resilience to disasters, resource and financial losses due to disasters were negatively associated with income levels.

Some older people, despite having limited access to credit markets, borrowed money from various sources such as banks, NGOs and moneylenders to restore their destroyed houses or build a new one, cover medical expenses, buy food, clothing, agricultural and fishing equipment and produce crops and fish. The older people and their households received both material and non-material support from various sources, but it was not enough to cover their living expenses. Similar observations were made in previous studies on the impact of Cyclone AILA in 2009, when relief efforts proved insufficient to absorb the setbacks. In particular, poor households had to borrow money from formal and informal institutions such as microfinance institutions, moneylenders or through other channels (Mallick, 2014; Subhani & Ahmad, 2019). The loss of employment and income, the rise in prices and the loss of purchasing power and credit liabilities have left the older people of the area in a state of vulnerability.

In addition to the loss of employment and income, older people and their households lost land, fishing boats, nets and machinery, production facilities, small business stocks and self-employment resources. The severe cyclones and floods damaged their fishing ponds and fishing grounds, and they lost their family's livestock and birds.

The dwellings of older people collapsed due to cyclones and floods in the coastal region. These dwellings were usually made of wooden and bamboo poles and corrugated iron sheets, which were not strong enough to protect them from the cyclone. Extreme weather events destroyed the houses of older people and their households, worsening their already vulnerable condition. As argued by Alam and Collins (2010), weak house structures and scattered settlements along the river lead to housing-related vulnerability among older people and their households in the region.

As the dwellings of the elderly could not withstand the cyclone, they had to seek safety in cyclone shelters. The journey to the shelter was hard for the older people, particularly for those who lived in remote areas, as they needed physical strength to negotiate the uneven roads and paths. Older people with arthritic knees and limited vision who do not have supportive people nearby become more vulnerable, unable to obtain food or water or feel at risk and are sometimes overlooked by volunteers and neighbours (WHO, 2008). In the current study, older people needed the help of other people to get to the cyclone shelter and they were sometimes injured by the strong winds and water. Despite having access to information about cyclones, older people were more vulnerable than other adults because their limited mobility often prevents them from going to cyclone shelters or other safe places, making them dependent on the help of others. Further, cyclone shelters were not big enough to hold all the people living in the area, and most shelters were far from the homes of older people. The structural conditions, atmosphere and facilities of the cyclone shelters were not suitable or adequate for older people.

It was evident that the most vulnerable group in society was older people. They were considered more at risk of death and disability and had less access to reconstruction assistance during and after the storm events (Gutman & Yon, 2014; Haq & Gutman, 2020; Kwan & Walsh, 2017). They needed more care and higher priority during and after extreme weather events to secure their lives and reduce casualties. The experiences of older people in the coastal region of Bangladesh show that extreme weather events had huge negative impacts on the lives and livelihoods of older people in the coastal region of Bangladesh. The climate events, directly and indirectly, affected the income, employment, expenditure and credit liabilities in housing, assets, food and nutrition, water and sanitation, production and health of older people and their households. Extreme weather events were found to continuously reduce the adaptive capacity of older people and their households, affecting their capital assets (physical, natural, financial, human and social capital) each year, which was exacerbated by existing socio-economic factors.

References

Alam, E., & Collins, A. E. (2010). Cyclone disaster vulnerability and response experiences in coastal Bangladesh. *Disasters, 34*(4), 931–954.

Burkart, K., & Kinney, P. L. (2017). Extreme weather and health. *Annual Review of Public Health, 38*, 103–124.

Eeuwijk, P. (2006). Old-age vulnerability, ill-health and care support in urban areas of Indonesia. *Ageing and Society., 26*, 61–80.

Government of Bangladesh (GoB). (2008). Cyclone Sidr in Bangladesh: Damage, loss and needs assessment for disaster recovery and reconstruction. Dhaka: Government of Bangladesh. Retrieved October 20, 2020, from https://modmr.portal.gov.bd/sites/default/files/files/modmr.portal.gov.bd/publications/83ccdd3b_1e9f_4e43_9583_cafeed377d73/AssessmentReport_CycloneSidr_Bangladesh_2008(Worldbank).pdf

Gutman, G., & Yon, Y. (2014). Elder abuse and neglect in disasters: Types, prevalence and research gaps. *International Journal of Disaster Risk Reduction, 10*, 38–47.

Haq, G., & Gutman, G. (2020). Climate resilience and older people. In D. Gu & M. Dupre (Eds.), *Encyclopedia of gerontology and population aging*. Springer.

Islam, M. R. (2018). Climate change, natural disasters and socio-economic livelihood vulnerabilities: Migration decision among the Char land people in Bangladesh. *Social Indicators Research, 136*(2), 575–593.

Islam, R., & Walkerden, G. (2014). How bonding and bridging networks contribute to disaster resilience and recovery on the Bangladeshi coast. *International Journal of Disaster Risk Reduction, 10*, 281–291.

Jahan, S., & Haque, A. (2016). Regional impact of cyclone Sidr in Bangladesh: A multi-sector analysis. *International Journal of Disaster Risk Science, 7*(3), 312–327.

Kwan, C., & Walsh, C. A. (2017). Seniors' disaster resilience: A scoping review of the literature. *International Journal of Disaster Risk Reduction, 25*, 259–273.

Mallick, B. (2014). Cyclone-induced migration in southwest coastal Bangladesh. *ASIEN, 130*, 60–81.

Mallick, B., Ahmed, B., & Vogt, J. (2017). Living with the risks of cyclone disasters in the southwestern coastal region of Bangladesh. *Environments, 4*(1), 13.

Masum, J. H. (2019). *Climatic hazards in Bangladesh: A literature review*. Coastal Development Partnership (CDP).

Paul, B. K. (2015). *Disaster vulnerability and resilience in Bangladesh.* Springer.

Paul, B. K., & Rashid, H. (2016). *Climatic hazards in coastal Bangladesh: Non-structural and structural solutions.* Elsevier. https://shop.elsevier.com/books/climatic-hazardsin-coastal-bangladesh/paul/978-0-12-805276-1 ISBN: 978-0-12-805276-1.

Saha, C. K. (2015). Dynamics of disaster-induced risk in southwestern coastal Bangladesh: An analysis on tropical cyclone AILA 2009. *Natural Hazards, 75*(1), 727–754.

Subhani, R., & Ahmad, M. M. (2019). Socio-economic impacts of cyclone AILA on migrant and non-migrant households in the southwestern coastal areas of Bangladesh. *Geosciences, 9*(11), 482.

Westlund, L., Poulain, F., Bage, H., & van Anrooy, R. (2007). *Disaster response and risk management in the fisheries sector.* Food and Agriculture Organisation of the United Nations (FAO) Fisheries Technical Paper No. 493.

World Health Organisation. (2008). *Older people in emergencies: Considerations for action and policy development.* World Health Organisation.

Chapter 6
Food Security and Health Vulnerabilities of Older People

6.1 Introduction

The vulnerability of older people to extreme weather events is multidimensional – geophysical, personal, and socioeconomic – that they face each year due to their individual and socioeconomic conditions and needs. The interactions between extreme climatic events and exposure, impact, socioeconomic conditions and capacities collectively influence the extent of their vulnerability. They suffer from extreme weather-events-induced loss of life and property, poor health and disease, unemployment and lack of income (as described in chapter five), which further increases their vulnerability. The devastating impact of extreme weather events on the health and socioeconomic conditions of older people results in deaths, injuries and a range of losses (Bayraktar & Dal Yilmaz, 2018). This chapter examines food insecurity and deterioration of health and wellbeing, which older people in the area experienced before, during and after extreme weather events. The findings showed that older people in the region were vulnerable to food insecurity. They had only limited access to family food and nutrients compared to their younger counterparts due to the continuous impact of extreme weather events on their socio-economic conditions. Their testimonies suggested that they suffer from various illnesses – physical and mental – resulting from extreme natural disasters. The extreme events were also found to affect their interpersonal and social relationships and mutual support, increasing their vulnerability.

6.2 Vulnerability to Food Insecurity

Extreme weather events increase the likelihood that older people and their households will go hungry and exacerbate the underlying causes of food insecurity. The impact of extreme events on food and nutrition varies among older people and their

M. I. Ali et al., *Extreme Weather Events and Vulnerability of Older People*,
https://doi.org/10.1007/978-3-032-11256-9_6

households in terms of socio-economic conditions, production capacity, availability of food, access to food and usability of food. Loss of food and livestock production and financial incapacity negatively affect food access and dietary diversity. The impact of extreme weather also affects the stability of the food supply during and after the disaster. Understanding the specific impacts of extreme weather events on food security is challenging because food vulnerabilities are widespread and highly contextual (Krishnamurthy et al., 2014). These vulnerabilities also depend on the capabilities of older people and their households, as well as the socio-economic support available in the region.

6.2.1 Inadequate Access to Nutritious Food

Inadequate access to food and nutrition has a harmful impact on the lives, livelihoods and wellbeing of older people, limiting their ability to work, engage in daily activities and live independently. Food vulnerability limits their capacity to endure and cope with the impacts of extreme weather events (FRAC, 2019). The lack of cultivable land and resources affects the availability of food and nutrition for older people in the region, as they have less food available due to lower production capacity. The extreme weather events that occur every year limit their socio-economic capacity to manage daily food. It has been observed that most of the elderly suffer from malnutrition due to a lack of food in the family. The food intake of older people depends on the availability and accessibility of food in the family. Some older people in the region find it difficult to obtain their daily food. Older people living in their sons' or daughters' families receive less food. This is because their sons and daughters first think more about their children than their parents. Most older people in the area have little or no access to nutritious food such as meat, milk, fish, eggs and fruits. Chandresh again said:

> I eat some of it [nutritious food], but it is rare. Where am I going to get money to buy these foods? The price of a dozen eggs is BDT100 and a litre of milk costs BDT60. The price of meat is high for us; it is BDT500 to 700 per kilo. [...] Most of the time I cannot eat nutritious food. If I buy this kind of food, I cannot buy rice or curry dishes. I usually eat meat once a month when we earn more. We cannot eat fruits, but I have some coconut trees from which I get some coconuts.

Fakir Shah from Padma village and Shuvasm from Haritana village similarly reported that they hardly get nutritious food because of their financial incapacity. They said that they have to wait for months or even years to get meat, milk or fruits. Most older people usually eat rice, curries, potatoes, vegetables, dal, and bread. Rashidul from Haritana village eats a small amount of food at different times of the day. He eats rice only at night. During the day, he takes some tea, biscuits and dried rice. However, as he lacks financial resources, he cannot get nutritious food. Eighty-year-old Abdullah gets a meal twice a day - at noon and in the afternoon. In the morning, he drinks water and has some tea. Sometimes he also eats rice and bread with curry in the morning. "I have two to four eggs a month, there is no fruit or meat. How can I eat and with what?" - he said. He eats fish when someone gives him a few pieces or when he can catch fish. Alfaz believes he needs nutritious food for his body, but he cannot buy fish, milk, meat or eggs.

In general, older people who are vulnerable to nutritional stress have a lower physical reserve, which limits their ability to recover in the face of an acute health threat or stressor from extreme events (Porter Starr et al., 2015). Some older women who had lost their spouses were unable to access food and nutrients. "I do not remember eating any fruits after my husband died," said Palkey, a widow from Ruhita village. She said she does not get milk, eggs or other nutritious food.

In another study, it was found that the majority of the elderly had an insufficiently varied diet (four or fewer out of ten food groups). They were found to be undernourished and at risk of malnutrition (Mridha et al., 2021). The same is true for Amena, who said that she does not eat as much as she used to. She eats rice or bread once or twice a day. She is too poor to spend on nutritious food. She added:

> How then am I to eat [nutritious food], O father? Does not it take the money [to buy nutritious food]? It costs money. My daughter sometimes sends me fruit and biscuits through my grandson. I eat fish, but I do not buy it. I can only eat fish if my daughter can catch it. I cannot buy and eat meat. I ate meat many days ago, but I did not buy it. I ate some beef at one of my brothers' houses and chicken at another.

The experiences of older people provide insight into the issue of access to food and nutrition in the region. Most of them said that they eat rice one to three times a day, sometimes a piece of bread or biscuits. They cannot eat nutritious food because they do not have money. In addition, it may be more difficult for older people to eat healthy food because of their age. This can reduce their physical reserves and limit their ability to recover in the face of a serious health problem or stressor. The physiological and sociocultural factors that affect food intake and behaviour are also important, and older men may be more vulnerable to malnutrition (Porter Starr et al., 2015; Thompson et al., 2017).

6.2.2 Scarcity of Food

Most older people in the study area reported that they had suffered from food shortages in the past after extreme weather events. Through the loss of crops and gardens, household stocks of food, properties, livelihoods and purchasing power for food, extreme events weaken households' potential for adequate food supply and access to nutrients (GoB, 2008). Many participants shared their experiences of food scarcity after the cyclone and floods. Many older people and their family members (girls, young children and pregnant and lactating women) were vulnerable to malnutrition and micronutrient deficiencies during extreme weather events. Although they needed comparatively nutrient-rich food, they could not share their food equitably among themselves and sometimes went without food. An elderly person, Rashidul of Western Haritana village, aged 80, shared the following:

> During the past disasters, I had problems managing food for the family. During the last cyclones, I was without food for two to three days. At that time, I bought bread, which cost BDT 5 in the shop, in twos and kept myself alive by eating it with water.

She further said that they had little or no food to eat after the cyclones. Older people claimed that they only had water and a piece of bread or a biscuit for the whole day after the disaster. An elderly fisherman, Shamiul, from West Hatitana village described his miserable experience when I asked him about the post-cyclone food crisis:

> Do not say that sir, I did not eat for many days. For days I had only water from panta bhat [a dish traditionally made from leftover cooked rice soaked overnight in water, usually served in the morning with salt, onions and chilli], which I had brought for me from their home. [...] I had been drinking only water for days.

Most of the older people said that the availability of food was inadequate after the cyclones. They also said that they were unable to buy food due to a lack of money, loss of livelihood, destruction of food shops and unaffordable food prices. As a result, they sometimes had to look for wild food from the forest to survive after extreme climatic events. Chandresh from Haritana village described lack of access to food after the cyclone thus:

> We had problems getting food during the disaster. If we did not get relief, sometimes we had nothing to eat for days. Two years ago [after a cyclone], it was very difficult to deal with food during a disaster. There was no more cooking. Then life depended completely on aid. If we did not get food aid, we had to rely on nature. We collected lilies, taro leaves and various wild plants. In one disaster, I had nothing to eat for two days, nothing – salt, rice, fire, water, or curry – was available. The cyclone destroyed everything. No one gave anything away to others.

Some said that they had to wait for supplies from relief agencies and most of the food provided by the government and NGOs was dried goods. It was very difficult to get cooked food immediately after the cyclone. Some older people living in remote areas said they were starving because they did not receive food in time. As Hossain narrated:

> During the disaster, we ate puffed rice or flattened rice. Then we had to eat once a day [or] we had to go without food for one or two days. When the cyclone started, it was not possible to eat anything until [the end of the cyclone was reached and] relief was given. For example, I took shelter at the hospital of Upazila headquarters [before] the cyclone AILA; the Upazila Nirbahi Officer gave us some biscuits during the last cyclone. (Hossain, 67 years old, Badurtala village)

People were generally unable to get food from the onset of the cyclone. They had to wait for food deliveries or somehow fend for themselves. Some reported that they were hungry when the cyclone started but they only received food after the cyclone ended. Fakirshah described, "I was starving at the time of the disaster. I had eaten at noon on the day of the cyclone and ate again in the evening of the next day [when the cyclone ended]."

Most reported that relief goods provided by the government and NGOs could not be delivered directly to the victims after the cyclone because various hazards, such as uprooted trees on the roads and paths, brought the delivery system to a standstill. As a result, the elderly had to rely on what was nearby. One of the participants, Sabdar 66 from Padma village, explained:

> First, when there was a disaster, there was problem getting water. Second, vegetables, fish and meat were not available. At that time, mainly pulses and potatoes were eaten. [...] Aid could not come by road as the cyclone had uprooted many trees and blocked the road.

It sometimes took a day or two to get food after the event, and then the residents who were not affected by the Cyclone or NGOs distributed relief goods. However, the relief goods were insufficient. Some elderly people said that a small amount of rice was also provided by government organisations. They had to manage in these conditions for almost 15 days before they could buy food with their earnings or with borrowed money from banks or NGOs. An 80-year-old participant, Enam, from Padma village explained:

> At that time there was no food at home, what were we supposed to eat? We had to go hungry during [and sometimes after] the floods and cyclones. Let us say I went hungry for two days during cyclone SIDR. After that situation, the NGOs provided food. Both the government and the NGOs provided food. [...] The food I got was not enough; we could not manage what they distributed. It was only enough for one or two days. Then we had to meet the expenses [for buying food] through our work. Otherwise, we [sometimes] had to sell our properties or possessions.

During and after the extreme weather events, there was another problem with the food supply. It was common for elderly people to wait for days for food, as the delivery and distribution of relief goods was a lengthy process. Abul, 70, from Padma village, explains the situation:

> It was getting late to get aid. People did not get it on the first go. It took time to organise it. It took almost 15 days for the officials to get the necessary papers. Within those 15 days, we got into trouble. We did not get food aid immediately after the cyclone subsided. The boat with the aid came, it was unloaded, and then [our] name was registered. Then the guard called the members and the chairperson. This process took almost 15 days.

For most of the elderly, work was the only way to support themselves or buy food after supplies from various sources ran out. However, this depended on whether there was work for the older people in the area. Some of them expressed the need to take temporary casual jobs for cash payment or for food at substandard wages. Abul said:

> At that time, there was great suffering for food. Work was not available at that time. [...] No one gave to others because they were also poor. No one gave even if they begged. How could they give? They had nothing else! So, we had to work as casual labourers for food. Cyclone piled up dirt in people's houses. If I could clean it up for someone, they would give me BDT100 with which to buy some food.

Again, food was not sent to the houses of the elderly, but they had to go to the distribution centres or cyclone shelters to get food. This was very difficult for some older people because of their physical condition. Elderly people who had sought shelter earlier and stayed at the cyclone centre only received some dry food as emergency assistance during the cyclone at night. Those who had not gone to the shelter during the cyclone said that they did not receive any food as they stayed at home. This was because the food was not distributed door-to-door but at specific locations. Some said that food shortages were a common outcome after any extreme weather event and that almost everyone suffered from these circumstances. An elderly person, Muhibul, from Badurtala village said: "If there was a cyclone, there would be food problems in the area. Not only I but everyone was affected by this problem". He also said that in the past, he had nothing to eat during a cyclone. "... It was about three/four days during and after the cyclone, and then I managed to eat!"

It is clear from the experiences of older people in the coastal region of Bangladesh that the unavailability and inaccessibility of food was a common situation after any extreme weather event in the region. Most narratives of suffering during the cyclone were intensive. Some were reluctant to disclose their suffering; they believed that it was not appropriate to tell their story. One elderly woman tearfully recounted her experience:

> [...] I had not been fed many times. Would it be beneficial if I told you? There was a feeling of great sadness in my life. During and after the immense cyclone and floods, I stayed three/four times. Together with my daughters, I sought shelter on the embankment by swimming during a cyclone. I cannot tell you how much I suffered.

People in the coastal zone were affected by food insecurity after the natural disaster, as the availability of food, accessibility to it and their consumption patterns were severely reduced. In addition to food shortages, there was also a shortage of drinking water after the cyclone. Extreme weather events significantly alter the salinity of water, especially in the southwestern coastal region of Bangladesh, leading to drinking water shortages (Ismail, 2016). Lack of potable drinking water was a common problem among older people, especially those who lived alone after the cyclones and floods. Most families of older people did not have their own tube wells. They had to fetch water from remote ponds, from others' tube wells or from the public water filtration system. One older person commented:

> For drinking water, a public filter had been set up on the bank of the pond half a kilometre away from my house. I drank the water by fetching it from the pond. Sometimes, when the filter was damaged, we had to drink the pond water by mixing it with alum [Aluminium Sulphate]. (Fakirshah, Padma village)

Each cyclone and flood exacerbated the water shortage in the region. Ponds overflowed with saltwater carried by the storm surge. Cyclones destroyed the public filtration system and tube well, sometimes leaving the elderly to rely on alternative sources of water. An elderly farmer, Hossain, 67, from Charlathimara village, explained:

> We could not even drink water because it was salty and polluted. We drank water from green coconuts or bought water from the shops in the Upazila headquarters, Patharghata or had to wait for volunteer help. We could eat only after getting help. Till then, one had to stay under the tree.

The shortage of food, water and other resources after extreme weather events forced those affected into a competitive and discriminatory environment. Some of the older people felt that they were the victims of discrimination in the distribution of relief and assistance by the government and NGOs. Sometimes they received no assistance at all and sometimes only by bribing the providers. Haulader, 70, a casual labourer from Padma village, said:

> I never got any help. I was selected for help, and I thought I would get along with others. But one person bribed a member and got it; I have not gotten it to date. It was distributed to others through bribery. I did not have money to bribe; I could not get relief; without a bribe, I got nothing.

The acute food problems continued and there was little or no food left in the homes of many older people and their families. They believed that discrimination against them continued even after the extreme weather events. Enam, 80, a widower from Padma village, expressed similar sentiments: "What would we eat? Only the powerful took the food that the government distributed. The persuaders took the food that was allocated to all".

The testimonies suggest that older people and their households suffer greatly from food insecurity before, during and after extreme weather events because food availability and accessibility are affected by the loss of production and disruption of proper supply and distribution. Cyclone SIDR, which hit Bangladesh in 2007, destroyed nearly 95% of crops in coastal districts (Asian Development Bank, 2013). According to the World Risk Index, disaster vulnerability and the risk of damage from extreme weather events such as floods and cyclones are determined by the spatial and temporal exposure to natural hazards, vulnerability, coping capacities and adaptive capacity of older people (Garschagen & Schmidt, 2015). Food insecurity negatively affects the ability of older people to cope with natural hazards and crises. There is an interaction between food insecurity and older people's vulnerability to extreme weather events and disaster risks. Hunger and food insecurity have an adverse impact on disaster risk because they significantly increase the vulnerability of the elderly population to extreme natural hazards (Garschagen & Schmidt, 2015).

6.3 Deterioration of Health and Wellbeing

The most common diseases affecting older people in Bangladesh include cardiovascular disorders, diabetes, arthritis, osteoporosis, skeletal deformities, obesity, behavioural changes, sleep disorders and dementia (Islam & Rahman, 2017). People aged over 50 often suffer from cancers such as prostate, lung, stomach, colon, breast and cervical cancers. More than 50% of people over 65 suffer from arthritis, which can cause pain and lower quality of life in rural and coastal areas (Islam & Rahman, 2017; WHO, 2018). These make it difficult for them to be strong both physically and mentally to face the challenges of extreme weather events and their vulnerable socio-economic conditions.

6.3.1 Physical Health Vulnerability

Older people in the coastal region find it difficult to move from one place to another and stay at home. They suffer from a lack of physical strength and are unable to move long distances due to various health complications. Some said that they could somehow walk to a nearby tea stall or grocery shop. An older resident shared his daily activities:

> I used to walk to the shop near my house once in the morning and once in the evening and have tea there. [...] I used to participate in various cultural activities, but now I cannot go because I cannot move in winter and summer [in extreme weather]. (77-year-old widower, Chandresh, Haritana village)

Physical weakness is one of the health vulnerabilities of older people in the region. Due to the decline in their physical strength, some are no longer able to perform daily activities such as bathing, washing clothes, etc. without the help of other family members. They are also no longer able to participate in various social activities as they used to in the past. For example, Rashidul from Haritana village can no longer do his daily activities without the help of others. He said:

> Normally I can go about my work, but when I take a bath, I have problems; I cannot bathe in the pond. I cannot wash my clothes. My wife helps me to get water for a bath and wash my clothes.

He also cannot participate as actively in social life as he used to. Sometimes he goes to religious ceremonies but not to cultural programmes. Again, chronic health problems are one of the most vulnerable conditions for older people in the region. Most suffer from various critical health conditions such as cardiovascular disease, kidney disease, digestive disorders, stomach problems and chronic pain. Further, they or their adult children cannot afford the medicines prescribed by the doctor to cure their diseases. Alfaz, a 66-year-old man suffering from pain, said: "I have many problems in my body now. I have pain all over my body now. I cannot do any work. I cannot carry the load. I do not have money, so I cannot take medicine".

Arthritis and pain are common health problems in older people that hinder daily physical activities and increase dependence on others. Raizuddin, a 69-year-old man, also reported that chronic pain in various parts of his body is the main enemy of his inability to move. He narrated:

> I am no longer hale and hearty. I am suffering from various diseases. I have been suffering from acute headaches for about two to three years. I have to take tablets [medicines] every day - one red and one white. These two tablets cost me six Taka.

Older women, especially widows who have to work longer hours to support their families, are more likely to suffer from physical problems. Rajina, a 63-year-old tea vendor from Ruhita Battala Bazar, like Raizuddin of Ruhita village, suffers from poor health and has pain in various parts of her body. She said, "I am in poor health condition. I have pain in my waist, knees, wrists and chest. I also have a bone problem." Sakimun, a 64-year-old woman from Haritana village, also suffers from chronic pain in her wrists and waist. Her physical problems make it impossible for her to do any kind of work. She said:

> [Oh] my son, I am burdened with so many diseases. I am on the verge of fainting because I have pain in my waist. My stomach feels like it is intertwining with my waist. Then I could not hold still.

Some reported that diseases like diarrhoea, respiratory infections, eye infections, various skin diseases and fever are increasing among the older people in the area. Chandresh, described his health during the cyclone:

> My health condition was very bad during the cyclone. I could not move around. The wind was cool, which caused breathing problems. I got a cold and cough [during the past cyclone].

Almost every older person suffered from health problems depending on how well they could bear the adverse conditions during and after the events. For some the cool wind that blew across the area during the cyclones caused colds, flu and fever after the events. For example, an older person said, "I caught cold after the cyclone because there was a cool wind then. I could not go to work". Some older people suffered from chronic illnesses such as flu and cough, heart disease and skin problems, which increased during the cyclone. To make matters worse, the extreme weather events destroyed medical facilities, especially local pharmacies in the area. As a result, older people had to wait for someone to attend to them or help them buy medicines at the district headquarters, which was five to 10 kilometres away from the study area.

Every year millions of people suffer from diarrhoea, skin diseases, mental disorders, malaria and dengue fever (World Bank, 2000). People in the coastal regions suffered from various chronic diseases such as heart problems, diabetes, and arthritis which can worsen during and after extreme weather events, leading to complications and even death.

6.3.2 Psychological Trauma

Mental health problems were found to be another health vulnerability alongside physical health problems among older people in the region. The losses and destruction caused by extreme weather events make them fearful of losses, injuries and even death every year. Most participants suffered from distressing incidents that occurred during cyclones. Rashidul from Western Haritana village said, "At the time of the cyclone, I got scared. There was such panic about floods and collapse of the dam." Another added:

> [my] body became weak during the disaster because I was afraid of cyclones [in the past]. Because in the last disasters, everything was washed away; only the foundation of my house was left standing. We fell into fear [when we heard the news of the cyclone]. (Hossain, 67, Badurtala village)

Mental health problems were common among older people and could cause worse symptoms during extreme weather events due to feelings of hopelessness and little or no support. They get scared when they hear the news about extreme weather events like cyclones and floods and remember their sufferings and worries from previous experiences. They reported tying down everything - beds, belongings - after hearing that a cyclone was brewing. As soon as the water reached the dam, people started crying. They went on to say that the bitter wind during the cyclone made their hands and feet numb; they struggled to grasp anything to react quickly and move. One of the participants explained:

> I was scared and upset. Everything had been broken and destroyed. I did not know where to go or what to do. It was shocking and puzzling. I tied everything together with a rope when I heard there was going to be a cyclone. I did not know which side to go to. As soon as the water reached the dam, the crying started. (Shamiul, 72, western village of Haritana)

Many older people were afraid and frustrated when they remembered their past experiences. They had witnessed the devastation caused by cyclones and floods in the past. After hearing the news of the cyclones, they did not know where to go for safety and what to do. They said it was terrifying and confusing. Raizuddin, a 69-year-old man from Ruhita village, commented, "I was scared during the cyclone. I called on the name of Allah and prepared to go to the shelter centre". Other elderly people also shared their feelings:

> I am scared of cyclones and floods. I go to the shelter centre. The place where the river has eroded is very close to my house. It is an eight-foot high embankment that can be eroded at any time by the striking force of the water. I live there in fear. I have not lived at home because it was flooded during the last cyclone SIDR. I took shelter on the embankment and held on to the trees. (Alfaz, 60, Padma village)

Rizeeya, 70, from Purbo Badurtala village, said he was scared during the cyclone and flooding and was thinking about where to go. He said many people drowned on the road (with a water grave) on the way to the shelter centre. Panicking during cyclones and floods was a serious problem among older people in the region. Most older people lived outside the dam, and some lived near the weak point of the dam or near the focal point of riverbank erosion, which caused greater anxiety. They had no choice but to live in fear as there was no other safe place to go to. Most did not stay at home after hearing the news about the cyclones because the dam was full during the last SIDR cyclone. They often left their house and took shelter at the dam by clinging to trees, believing that they would survive even if the house blew up.

Participants also recalled the past loss of their assets and circumstances. Based on their past experiences, they knew that their property and house would be lost forever, which made the situation even more worrying. An elderly person from Western Haritana village recalled an event from the past:

> During the last cyclone, trees had fallen on my house. When I noticed this after the cyclone, I fainted. I got scared when people announced that the cyclones would come again. [...] I did not stay in [my] house. I could not stay standing at the time of the cyclones. [...] When I hear the signal of the cyclone, I immediately go to my brother-in-law's house, which is made of bricks. (Rasidul, 80, West Haritana village)

The older people and their households believed that similar destruction could occur, which increased their fears. For example, "It [house and dam] would collapse if there were cyclones and floods again. I was totally scared about that [cyclone]. Everything [house, dam, crops] would be destroyed if the water from the sea hit again". The health problems related to the disaster were a growing concern and a direct consequence of the extreme weather events. In addition, there was the destruction or forced relocation of familiar, emotionally or culturally valued habitats, as well as indirect consequences, e.g. physical health problems and/or social and economic disruption to families (Berry et al., 2010).

Studies have found that deteriorating physical and mental health, frequent falls and limited care can make older people vulnerable (Barbosa et al., 2019; Brocklehurst & Laurenson, 2008; Sarvimäki & Stenbock-Hult, 2016). To build adaptive capacity to cope with socioeconomic hardships and extreme climate change, older adults need to be physically and mentally strong and know their limits. However, the experiences of older people in coastal Bangladesh show that the weaker physical strength and reduced mental capacity of older people increase their vulnerability to extreme events. Their poor health conditions may lead to the need to evacuate them in emergencies during extreme weather events. The lack of primary health care and limited access to health facilities also increase the vulnerability of older people in the region (Malak et al., 2020).

6.4 Discussion and Conclusion

Older people and their households in coastal Bangladesh face different types of vulnerabilities because of extreme weather conditions every year. Focusing on climate change and seasonality, Shams and Shohel (2016) analysed the existing threat to food security and livelihoods in the coastal region of Bangladesh. Their study showed that there were recurrent food shortages during the sowing and harvesting seasons. While the food deficit remained high (68%–95%) from mid-August to the end of October, it reached its lowest level (21 %) in December. The experiences of older people in the coastal region of Bangladesh show that food insecurity or vulnerability is common among the elderly population following extreme weather events. The food intake of older people in the study area is influenced by their ability to buy food or by the availability and accessibility of food within the family. What they get one to three times a day is what they eat, i.e. nothing special is prepared for them. There are days when they get nothing to eat at all. Nutritious food such as meat, fish, eggs and fruits are only taken when they are given by a close relative, a neighbour or at a sacrificial festival of the Islamic religion during the year. They cannot afford to buy them themselves. It was found in another study that rice, wheat, pulses and vegetables were the most commonly consumed foods, while fruits and sweets were the least consumed by older people (Quddus & Bauer, 2014). Further, a quarter of the elderly ate fish, while two-thirds of them ate meat for less than a week. It was found that food insecurity was inversely related to household income and education of the elderly and positively related to the number of family members. In addition, older people are unable to feed themselves regularly and their families essentially live on subsistence. Several studies have found that the majority of older people have incurred debt for subsistence.

This study showed that food and nutritional shortages increased following extreme weather events, especially cyclones and floods, which exacerbated food insecurity among older people. It has also been reported that past disasters were so extreme that affected households often had to sell their labour in advance at a very low price and mortgage their remaining assets at undervalued prices to buy food (Roy et al., 2009; Subhani & Ahmad, 2019). The current study identifies several causes of food insecurity following extreme events, such as unavailability of food because supply was affected by the destruction of communication infrastructures, loss of local production and access to food due to lack of income and savings, and price increases. Again, household cultural norms determine who has priority access to food. As shown in this study in households where food is scarce, older women and men often suffer from a shortage because they forgo food so that younger family members have enough to eat. In addition, access to food can be further affected because food distribution points are located in places that are difficult for older people to reach. These conditions mean that older people may not be cared for after cyclones and floods. A study of cyclone AILA in 2009 in the coastal region found that around 90 % of affected people had reduced food intake; for example, those affected had skipped meals once or twice a day (Subhani & Ahmad, 2019). In addition to food shortages, there was also a severe shortage of potable drinking water immediately after the cyclone devastated the region. The saline water contaminates surface water and damages tube wells, resulting in water vulnerability for older people and their households.

Many older people in the coastal region suffer from various chronic diseases, although health is their most important asset: it affects their ability to work and maintain a satisfactory standard of living (HelpAge, 2002). Most older people in the study area are vulnerable to diseases and infirmities. They are unable to get healthy and nutritious food and adequate medicines. They also depend on the local pharmacy to get medicines in case of illness, as health services in the area are inadequate. Their reduced physical strength and declining mental capacity were found to make them vulnerable to disease. Often, older people lack adequate sanitation and hygiene facilities, food security and family care and support, making them vulnerable to disease (Malak et al., 2020).

Older people in coastal Bangladesh experienced many physical and psychological problems during and after extreme weather events. They suffered from illnesses such as cold fever and flu, cough, skin irritation, diarrhoea, mental illness, malaria, abdominal disorders and dengue disease in addition to their existing illnesses. Extreme weather events severely affect the health of older people, causing numerous diseases (Kabir & Khan, 2016; Mian et al., 2019). Older people also suffered from waterborne diseases because they used contaminated, salty water from various sources. Further, the severe weather events destroyed medical care facilities, including local pharmacies and dispensaries, which exacerbated health vulnerabilities. Older people found it difficult to access health care due to mobility problems, distance from central medical services, lack of regional health networks and the disaster-related absence of health facilities for daily needs.

In addition to physical health problems, older people in the region reported mental health problems. They and their households were frightened when they recalled the suffering and grief they had experienced during previous extreme weather events. They were reported to have lost family members, jobs, homes, assets, farmland and livestock in the past, leading to mental illness among older coastal residents (Kabir & Khan, 2017). The reasons for this vulnerability are complex and vary depending on the type of disaster, the magnitude, the personal sense of loss and pre-existing negative factors such as comorbidity.

References

Asian Development Bank. (2013). *Implementation completion memorandum: Strengthening institutional capacity for social protection in Bangladesh* (Project No. 46284-001). https://www.adb.org/sites/default/files/project-documents/46284/46284-001-icm-en.pdf

Barbosa, K. T. F., Oliveira, F. M. R. L. D., & Fernandes, M. D. G. M. (2019). Vulnerability of the elderly: A conceptual analysis. *Revista Brasileira de Enfermagem, 72*, 337–344.

Bayraktar, N., & Dal Yilmaz, Ü. (2018). Vulnerability of elderly people in disasters: A systematic review. *Turkish Journal of Geriatrics/Türk Geriatri Dergisi, 21*(3), 467–482.

Berry, H. L., Bowen, K., & Kjellstrom, T. (2010). Climate change and mental health: A causal pathway. *International Journal of Public Health, 55*(2), 123–132.

Brocklehurst, H., & Laurenson, M. (2008). A concept analysis examining the vulnerability of older people. *British Journal of Nursing (Mark Allen Publishing), 17*(21), 1354–1357.

FRAC. (2019). *Hunger is a health issue for older adults: Food security, health, and the federal nutrition programmes.* Retrieved October 10, 2021, from https://frac.org/wp-content/uploads/hunger-is-a-health-issue-for-older-adults-1.pdf

Garschagen, M., & Schmidt, S. (2015). *World risk report 2015. Bündnis Entwicklung Hilft and UNU-EHS.* Retrieved October 10, 2021, from http://collections.unu.edu/view/UNU:3303#viewAttachments

GoB. (2008). *Cyclone Sidr in Bangladesh: Damage, loss and needs assessment for disaster recovery and reconstruction.* https://reliefweb.int/sites/reliefweb.int/files/resources/F2FDFF067EF49C8DC12574DC00455142-Full_Report.pdf

HelpAge International. (2002). *State of the world's older people 2002.* Retrieved September 6, 2006, from http://www.helpage.org/Resources/Policyreports

Islam, M. A., & Rahman, M. F. (2017). Health problems of elderly people in Bangladesh. *Journal of Armed Forces Medical College, Bangladesh, 13*(1), 1–2.

Ismail, H. (2016). *Climate change, food and water security in Bangladesh.* Future Directions International. Retrieved July 28, 2021, from https://www.futuredirections.org.au/publication/climate-change-food-water-security-bangladesh/

Kabir, R., & Khan, H. T. A. (2017). Study on the health status of coastal people in Bangladesh after cyclones Sidr and Aila. *European Scientific Journal, 13*(15), 10.

Kabir, R., & Khan, H. T. (2016). Climate change impact: The experience of the coastal areas of Bangladesh affected by cyclones Sidr and Aila. *Journal of Environmental and Public Health, 2016*, 9654753.

Krishnamurthy, P. K., Lewis, K., & Choularton, R. J. (2014). A methodological framework for rapidly assessing the impacts of climate risk on national-level food security through a vulnerability index. *Global Environmental Change, 25*, 121–132.

Malak, M. A., Sajib, A. M., Quader, M. A., & Anjum, H. (2020). "We are feeling older than our age": Vulnerability and adaptive strategies of aging people to cyclones in coastal Bangladesh. *International Journal of Disaster Risk Reduction, 48*, 101595.

Mian, I. A., Zulfiqar, M., Khan, N., Hanan, F., Qazi, M. A., Gul, N., Afsar, N., Jehan, S., & Fahad, S. (2019). Impacts of climate change on the health of older people in District Nowshera, Pakistan. *International Journal of African and Asian Studies, 57*, 2019.

Mridha, M. K., Hossain, M. M., Khan, M. S. A., Hanif, A. A. M., Hasan, M., Hossaine, M., & Mitra, D. K. (2021). Nutrition and health status of elderly people in Bangladesh: Evidence from a nationwide survey. *Current Developments in Nutrition, 5*(Suppl._2), 39–39.

Porter Starr, K. N., McDonald, S. R., & Bales, C. W. (2015). Nutritional vulnerability in older adults: A continuum of concerns. *Current Nutrition Reports, 4*(2), 176–184.

Quddus, M. A., & Bauer, S. (2014). Food security and morbidity of elderly in disadvantaged rural Bangladesh. *Journal of the Bangladesh Agricultural University, 12*(1), 95–104.

Roy, K., Kumar, U., Mehedi, H., Sultana, T., & Ershad, D. M. (2009). *Initial damage assessment report of cyclone AILA with focus on Khulna district*. Unnayan Onneshan-Humanity Watch-Nijera Kori.

Sarvimäki, A., & Stenbock-Hult, B. (2016). The meaning of vulnerability to older persons. *Nursing Ethics, 23*(4), 372–383.

Shams, S., & Shohel, M. M. C. (2016). Food security and livelihood in coastal area under increased salinity and frequent tidal surge. *Environment and Urbanization ASIA, 7*(1), 22–37.

Subhani, R., & Ahmad, M. M. (2019). Socio-economic impacts of cyclone AILA on migrant and non-migrant households in the southwestern coastal areas of Bangladesh. *Geosciences, 9*(11), 482.

Thompson, J., Tod, A., Bissell, P., & Bond, M. (2017). Understanding food vulnerability and health literacy in older bereaved men: A qualitative study. *Health Expectations, 20*(6), 1342–1349.

World Bank. (2000). *Bangladesh: Climate change and sustainable development* (Report No 21104 BD). World Bank.

World Health Organization. (2018). Ageing and health [Fact sheet]. Retrieved June 2022, from https://www.who.int/news-room/fact-sheets/detail/ageing-and-health

Chapter 7
Social Protection and Vulnerability of Older People

It shall be a fundamental responsibility of the State to attain, through planned economic growth, a constant increase of productive forces and a steady improvement in the material and cultural standard of living of dis of undeserved want arising from unemployment, illness or disablement, or suffered by widows or orphans or in old age, or in other such cases.
(Article 15(d), The Constitution of the People's Republic of Bangladesh)

7.1 Introduction

Extreme weather events induced by climate change have recently increased in frequency and intensity in the coastal region of Bangladesh (Ahmed, 2005; Kabir et al., 2016; Karim & Mimura, 2008). In addition, changing weather patterns will continue to, directly and indirectly, affect the lives and livelihoods of millions of people, as will changes in average temperatures, rainfall and other weather conditions. These extreme events are more likely to affect developing countries like Bangladesh because they are most exposed to climate-sensitive sectors such as agriculture, which account for a substantial portion of their economies and the livelihoods of the poor and vulnerable. Older people, like other vulnerable groups, are most affected in these countries as they are more exposed to extreme events and have lower adaptive capacities (Béné et al., 2014). The social protection system offers significant potential to protect poor and older people from their disaster risk and vulnerability and to facilitate future adaptation to extreme weather events (Davies et al., 2008).

This chapter discusses the role of social protection programmes in addressing the vulnerability of older people to the impacts of extreme weather events in the coastal region of Bangladesh. The extent to which social protection programmes help to reduce older people's vulnerability to extreme weather events depends on the availability and accessibility of these programmes in each community. In many cases, there is a gap between the scale and reach of formal social protection systems, and

M. I. Ali et al., *Extreme Weather Events and Vulnerability of Older People*,
https://doi.org/10.1007/978-3-032-11256-9_7

certain segments of the population are necessarily excluded despite their needs. In such situations, traditional community-based forms of social protection programmes can reduce risk and vulnerability within a community and fill some of the gaps left by formal measures (Norton et al., 2001). Because of this, all social protection systems, both formal and informal, are discussed in this chapter.

7.2 The Contribution of Formal Social Protection Programmes

Social protection programmes play a crucial role in reducing the vulnerability of older people to extreme weather events. Social protection can help to better address the structural causes of poverty and vulnerability to weather extremes to strengthen and transform the productive livelihoods of the poor and vulnerable (Tenzing, 2020). Older people who participated in the study discussed both formal and informal social protection programmes available and implemented in the region. Formal social protection refers to policies and programmes implemented by the state alone, with or without the help of other parties such as NGOs and donors, as well as organisations within or outside the state (Kasseeah & Tandrayen-Ragoobur, 2011). Almost all participants reported that they were able to cope with different situations after extreme weather events thanks to the various social protection programmes run by government agencies and NGOs. However, they felt that the government and its associated local bodies helped them more than NGOs. Nevertheless, they received help, more or less, from all kinds of organisations. One participant, Fakir Shah, from Padma village said, "... both the government agencies and NGOs are doing their best to help the vulnerable older people and their households. NGOs are trying to provide a fairly equal share for all." He further elaborated:

> After the cyclone, government officials and NGOs distributed chira [flat rice] and muri [puffed rice]. This went back and forth for two days. After that, the NGOs distributed some relief goods. The government had given a small amount of rice. We spent almost 14 days in this way. Then we repaired [our] houses by taking loans from NGOs.

Old Age Allowance, Widows Allowance, and other types of social protection programmes are examined here to explore how they mitigate the vulnerability of older people in the region during and after the weather extremes. Some older people were eligible for these types of formal social protection programmes, while others were not due to the means-testing of these programmes. In addition to the formal social protection programmes, some informal social programmes helped older people and their families to cope with the consequences of various extreme weather events. All participants, beneficiaries and non-beneficiaries of social protection programmes appreciated these programmes for providing better livelihoods for older people in the region.

7.2.1 Improvement in Living Standard

Social protection beneficiaries can improve their standard of living by buying more food, medicine and other daily necessities. A social protection system that provides low benefits to a large number of beneficiaries can reduce poverty more effectively than one that offers high benefits to a small number of beneficiaries (Begum, 2012). Social protection programmes can transform the lives of people in a community compared to those who do not receive such kinds of benefits. In most cases, beneficiaries report that their standard of living has improved significantly after receiving the allowance. Eighty-year-old Rashidul has been receiving the Old Age Allowance since the programme began in 1998. He receives the money from the bank on the scheduled day and uses it as needed. He said he does not share it with his wife, sons or daughters. In addition to rice, he buys medicines, curry and other items. He said that he also bought clothes–a saree–for his wife and gave some [money] to his grandchildren. He also saved BDT 50 from his allowance. He commented on the benefits of these programmes:

> I am a little better off now than before I received the allowance. If I do not receive it, the entire amount [the allowance he receives] is in deficit and I get into trouble. I can buy almost 30 kilograms of rice with the money. How am I supposed to buy this rice if I do not get the allowance?

Most of the older people indicated that they were almost entirely dependent on the allowance for their survival. Without the allowance, it would be almost impossible for them to support themselves and their families. Many also think like Chandresh, who states that without the Old Age Allowance, he would not be able to manage the necessities of his life and living. For example, Rashudul says:

> I will face problems if the government decides to cancel my Old Age Allowance. If I do not get any money, I will face difficulties. If I did not get an allowance, I would not be able to buy rice.

Moreover, even those who were not recipients of those benefits believed that the social protection programmes were beneficial to older people. They think that the allowance would enable them to meet their daily needs and help them to live more independently. Although Asadul is 80 years old, he does not receive the Old Age Allowance. However, he is aware of the value of the allowance and says, "If I got the Old Age Allowance, I could eat and live in peace." Razeeya expressed similar sentiments, saying that she does not receive any allowance because of her unknown face in the community. When she went to see the chairperson and members of the Union Council, they only told her that she would receive the allowance. However, she has not yet received it. Nevertheless, she realised the benefits of the allowance and said:

> If I got this allowance, I could take [some] medicines; I could also buy necessary food at the market. In a sense, I would be able to eat some healthy food. I want to stay here at home as long as I live. If I could fix it [the house], it would be better for me [to reside in].

Shuvasm, a village vendor from Haritana village, is 60 years old and does not receive the old age allowance; however, his mother is entitled to it. He said:

> My mother gets this allowance. It is very helpful. She can buy medicine and food with the money. But it is not enough for her to survive. We take care of my mother. She lives with us by rotation. [...] If I could get the old age allowance, that would be helpful for me. In my old age, it will help me to reduce my workload. Although I cannot provide for my family's entire living expenses, it comes in handy when I am unemployed. I would use it at times of disaster when I do not get any work. It helps me then to buy food and other things.

Fakir Shah, a 61-year-old small-scale fisherman, also does not receive the old age allowance or other support because he has not reached the eligible age for any programmes yet. However, one of his sons is disabled and receives a disability allowance. According to him, this allowance is useful in that he can buy clothes for his family with the money his disabled son receives every year. "Without the allowance, I would have faced a severe lack of money throughout the year to buy clothes for my family," he recalls. Similarly, Ismail, a 75-year-old marginal farmer, does not get the old age allowance yet. In his opinion, it would be beneficial for his family if he received Taka 3000 as an allowance every 6 months. He said, "I am hard up now. That is why I cannot buy anything. [If I get the allowance], I can buy everything and eat something." Like Ismail, Rahimon, a 65-year-old widow, said that it would have been beneficial for her if she had received the allowance: "My sons feed me. If I got an allowance, I would stay healthy by taking medicines. That would have been beneficial." Her experience shows that older people who do not receive any form of support need to rely on others in the family.

Social protection programmes have a significant impact on the capacity of older people and their households to cope effectively with a variety of extreme covariate events. Such programmes may also have helped to cushion the impact of rising food prices following extreme climatic events and protect them from hunger. Public works or government job creation programmes have significantly reduced the impact of the financial crisis during various events. In a study of the self-perceived impact of the government's 100-day employment guarantee programme, more than three-quarters of participants said the programme had improved their overall economic situation, food consumption and standard of living (Fiorillo & Matin, 2009). Various social protection programmes also helped older people and their households to avoid forced migration and provided them with alternative employment opportunities (Bastagli & Holmes, 2014; Fiorillo & Matin, 2009; Grosh et al., 2011).

7.2.2 Acceptance and Trust in Family and Community

Older people may feel safe in economic emergencies following extreme weather events if they are treated with respect when they turn to their family or community for emergency assistance. The various social protection programmes, particularly the Old Age Allowance and Widow Allowance, improve the dignity of older people in their families by making them more attached to their family members. Family members cared more for older people when they received certain allowances. Many older people indicated that social protection programmes gave them greater opportunities to actively participate in family decision-making. Many also felt that receiving an allowance increased the acceptance of beneficiaries by their families. A similar sentiment was expressed by Fakir Shah:

> My family members would take care of me a bit more if I received the old age allowance. The children knew that their father would get some money at the end of the month and that he would give us some money. Then they told their wives that they should respect our father. They also told their daughters that they should behave well towards their grandfather. Your grandfather gets an allowance from the government.

It was clear that the sons and daughters and other household members of benefit recipients behave well when they receive an allowance. Raima Begum, a widow of Western Badurtala Cluster Village, had similar views:

> [...] That is for sure. Now I have something, so everyone takes care of me. Otherwise, they would not take care of me. They would not allow me to stay here either. Rather, they would say, "Go away because your husband is dead". I get allowance money, and my disabled daughter also gets it. For this reason, everybody behaves well with me. I see that other widows in the village do not receive old age allowance; their sons and daughters do not ask them for anything. Like them, it might have happened to my lot as well.

She again said that her sons also ask her for everything, as she has some money. Raima Begum lives with her elder son and gives him money. She has realised how important the allowance is and says sadly:

> If the allowance stops, I will have to beg; I have no way without begging. I shall get some rice as alms from people. I shall take it in my bag and have to cook it to eat. There is no other way without the allowance.

Some said that the social protection programmes have helped increase the confidence of people in the community toward older people. Many reported that most people trust older people to borrow money and get necessary goods, as they assume that these people will pay back the dues and loans after they get their allowance. This was confirmed by Rashidul: "The shopkeepers give me food and other things, trusting that I will be able to pay back the money after getting the allowance". The shopkeepers also stated that they give necessary goods to older people on credit in the hope that they will pay when they receive the allowance money. A 60-year-old participant, Kamal, a shopkeeper from Badurtala Bazar, added:

> I have some customers who take their needed goods on credit and say they will pay back when they get the allowance money. Since I believe that, I give them products in advance on credit. They pay it after they receive their allowance. So, there is a firm faith in them [older people].

Lailunahar is a 60-year-old widow from Padma cluster village, who offers services such as sewing Kantha [kantha is a type of fabric used in Southeast Asia, especially in Bangladesh and the Indian states of West Bengal and Odisha, as an alternative to bed sheets and quilts] to people at home. This is an advantage for her as she receives the widow allowance. She says, "Shopkeepers give me goods without money because they know I will get the allowance and pay them back. The pharmacy gave me medicine without money when I asked for it." After she receives the allowance, she pays it back. This is how she describes the benefit of the allowance in her daily life:

> Ah! I must die if the government does this [stops the allowance]... Even if it is paid off after 6 months, we have the hope and assurance that I will pay back the debt from anywhere. How will we live if it is stopped now? What will we eat?

Social protection programme like Old Age Allowances in Bangladesh goes a long way in improving the social relations of the elderly beneficiaries. The relationship of recipients with their sons, daughters, relatives and neighbours improved significantly after receiving the allowance compared to before. In all cases, older people who had good social relationships before receiving the allowance were able to maintain them and those who had not good social relationships were able to improve them significantly. It appears that the Old Age Allowance scheme has a positive effect on the social relationships of older people in two ways: first, it prevents the deterioration of social relationships of those who used to have good social relationships, and second, it helps to improve the social relationships of those who had not good social relationships (Karim & Hossain, 2013). It is clear from the evidence that the beneficiaries who receive old age allowances are treated well by local grocers. During an emergency, grocery stores provide essentials to older people on credit. Therefore, older people feel secure believing that regardless of their financial situation, they will still be able to get emergency food and medicine at the drugstore or grocery store because they have some kind of money.

7.2.3 Saving, Investment and Loan Repayment

Extreme weather events have a profound impact on the savings and investment behaviour of older people and their poorer households, thereby increasing poverty and vulnerability. Some resilient households have access to savings in the form of cash and assets that can provide a buffer after a shock from climatic events. Social protection in the form of cash transfers, for example, provides recipients with an additional source of income that can facilitate precautionary measures such as savings, and investments following extreme events and support their adaptation (Bowen et al., 2020). In most cases, older people use their social protection benefits for food and medicines. However, some beneficiaries reported that they could save and invest part of their allowances that generated income. Some spent on buying chickens, ducks, and cattle, and that caring for these animals would help them after climatic events. They can sell these animals and birds in hard times that follow extreme weather events. They have invested the allowance money so that they can deal with future extreme events better. Muhibul from Padma village was one of them who said:

> It is beneficial for me. I used to work on my own. I saved what I got [as allowance]. I bought a goat and two ducklings [with the allowance] and raised them. I was able to take medicine and to buy food.

This shows that the old age allowance helped the beneficiary, Muhibul, to save and invest for future uncertainties. Some beneficiaries used a portion of their allowance to restore their house or repair the house's amenities following the cyclone. Fakir Shah stated, "... [my] house was repaired a little bit after I received the grant at the time of a cyclone and flooding." Some other beneficiaries mentioned that they bought materials to repair their house after the extreme weather events. Rashidul said, "I [once] bought some tin with the money I got as allowance." Sakimun, a 64-year-old widow from Ruhita village, also said she repaired her damaged veranda after a cyclone with the allowance money she first received along with some additional money from her daughters. These testimonies suggest that the different allowance programmes help older people and their families to prepare for extreme weather events.

Alamgir Hossain is not eligible yet for the old age allowance, but his wife receives another benefit - 30 kilos of rice per month. Alamgir believes that the pressure to work will decrease when he receives the old age allowance. He said that with this money [the allowance] he could do something for himself and his family members if need be. Debnath, a 67-year-old carpenter, receives the old age allowance but cannot keep it with him for several months. He said:

> How can I keep it when I am in debt? I cannot be free of debt. When I had no income, I used to purchase things in shops on credit. I buy groceries and daily necessities in the shop account and tell them that I will pay off my debts after receiving the old age allowance. The shopkeepers trust me and give me the products on credit.

For some people, the allowance they receive helps them repay loans they took from their neighbours or relatives during the crises. Kabiruddin, a 70-year-old resident of Padma village, said he had been receiving the allowance since 2007 before cyclone SIDR hit. The money was withdrawn from the bank at the scheduled time and spent on various things. He said run into debt many times. As a result, he had to pay the debt with the allowance. He said, "... after bringing the allowance it is spent sometimes on the spot. How many days does this go on? It is spent in one day. I had to pay back the shopkeepers' debt". Again, he added:

> We have reached old age. Sometimes the owner would not pay the wages after mowing the land all day without bathing and eating, and I would be in a rage. 3 thousand Taka [the allowance] is paid by the government as the old age allowance after 6 months. In these 6 months, I borrowed a lot of money from others. I am paying back the borrowed money. My son cannot maintain it on his own. [...] The shopkeepers give me the necessary products on credit. They said you would pay back the dues when you receive the allowance.

The various forms of social protection help many older people, most of whom are either self-employed or in the informal economy and do not have access to savings and loans, which affects their ability to invest in and diversify their livelihoods (Canelas & Niño-Zarazúa, 2022). Social protection can promote financial inclusion among older people, which can be critical to improving their ability to cope with, prepare for and adapt to extreme weather events. Access to financial services can help older people and their households reduce their vulnerability to shocks by facilitating saving to support preparedness; enhancing opportunities for income generation, asset accumulation and consequently adaptation (Barrientos, 2011; Bowen et al., 2020; Cull et al., 2014).

7.2.4 Health Care and Food Security

Older people are vulnerable to significant health problems - both physically and mentally - during extreme weather events. In addition, extreme weather events affect their food security by destroying production and livelihood opportunities. Social protection programmes help older people and their families to reduce health vulnerability and food insecurity during and after extreme weather events. Most participants reported that they had previously obtained some of their needed medicines with the help of family members, friends, and relatives. However, after receiving the allowance, they purchased their medicines. Chandresh from Haritana village has been receiving the old age allowance for almost 15 years. He is receiving BDT500 per month but every 6 months. He feels that the allowance contributes significantly to his life. He said:

> I can buy food, clothes and medicines for myself and sometimes something for my grandchildren and other family members. Sometimes I spend some money to fix my house. [...] it [the allowance] helps me to live a little better than before. Before, I could not afford to buy milk, but now I can. At least once a month I can afford milk. I can buy medicine for myself and some food for my grandchildren. [...] it is very difficult for me to live without an allowance. If the government does not give this money, I have to beg [from other people].

Social protection, especially cash transfers, reduces the level of poverty and vulnerability of recipient households, improves access to essential health services, and supports relationships within households (Lloyd-Sherlock, 2006). Social protection such as old age allowance has an impact not only on beneficiaries but also on members of older people's households. It is also believed that old age allowance has a positive impact on health-related quality of life as it enables beneficiaries to contribute financially to their households (Rana & Ahmed, 2010).

In addition, recipients of the allowance said that their daily dietary habits had changed, and they were able to eat more than before. The variety of foods they ate at each meal had also increased after receiving the allowance. That is, they could eat nutritious foods and have some freedom to buy necessities. Raima Begum shared the following:

> I receive the widow allowance. I receive it every 6 months. I draw my allowance and keep it for myself. My sons do not take it. They take it when I give it to them. I use the allowance to buy clothes or something to eat [buy food]. When I get the allowance, I can eat what I want, maybe some chicken and fish.

In general, most (60%) of the allowance received was spent on food, followed by health care (19%) (Haider & Mahamud, 2017). Some reported being able to afford other food items in addition to regular meals, including biscuits, cakes, tea and puffed rice. Abdullah from Charlathimara Village has been receiving old age allowance since the year after SIDR in 2007. Every 6 months, he says, he receives 3000 Taka (500 Taka per month). He spends the money he receives on rice, flour, chira, fried rice and medicine. He has to buy medicine frequently, especially for fever, pain and arthritis. However, he finds it difficult to meet his basic needs with this small amount of money. He said, "I have to pay the shopkeepers after I receive the allowance; I pay the dues. With the rest of the money, I can manage only a few days." Raizuddin was 69 years old and frequently suffered from various diseases. He received the old age allowance which enabled him to buy his medicine. He said:

> I spend everything [my allowance] on medicines and there is nothing left to buy rice and pulses [daal]. Sometimes I suffer from a fever. Whenever I go to work, I feel pain in my body. When I feel pain in my body, I have to take medicine bought with this money [allowance]. [...] but it does not last more than a month.

It is clear that the beneficiaries of the social protection programme can meet their minimum daily needs (food, clothing and others) and necessary medicines when they are sick. It can be said that the standard of life of older people has improved after receiving the old age allowance (Rana & Ahmed, 2010; Uddin, 2013). Some received the widows' allowance which helped them to support their families in economic crises. For example, Momina Begum has been receiving the Widow Allowance for 10 years. Initially, she received 125 Taka per month. Now she gets 5 hundred a month. She commented:

> I get a widow allowance. I draw my allowance and keep it for myself. I buy clothes for my daughters and sons and myself. I also buy daily necessities - rice and vegetables. Sometimes, I spend almost all my money on food.

Momina went on to say that her elder daughter used to work in a garment factory in Dhaka before she got the allowance. At that time, she [her daughter] used to give her a bag of rice and 1 thousand Taka per month. Now she [her daughter] could not give her any money because she got married. She explained in frustration, "[...] I will die if the authority stops giving the allowance. Without that money, I would not be able to buy clothes anymore." Amena Begum, a widow of Padma village, exclaimed, "Is there any other way than begging if the government stops paying the allowance?" This shows how important the allowance money is to the survival of many older people in the area. Lailunahar, a 60-year-old widow, has been receiving the allowance since the programme began. She spends the allowance on buying her necessary goods. "I can buy clothes when I have nothing to wear. If there is no rice at home, I can buy it. I buy goods and medicines to take home," she said.

It was noted that women who have been abandoned by their husbands or widowed can benefit from the Widow Allowance Scheme. They come out of severe food insecurity, receive better medical care, and can dress relatively well (Islam, 2007). The beneficiaries of various allowances were found to have greater access and control over household assets, their mobility and decision-making within the household and community. They also have time and opportunity to participate in various sociocultural events that deal with political activities, media use and information. They were able to make decisions about their health care and economic activities using their resources (Chandra et al., 2017).

7.3 Contribution of Informal Social Protection

Informal social arrangements to provide socioeconomic support are considered when formal social protection structures do not exist or are inadequate to provide adequate support to older people and their households (Oduro, 2010). Informal social protection is mainly non-governmental in nature. Informal social protection (ISP) is structured through social relationships. Social networks and relationships and mutual help play a significant role for older people and their households in reducing vulnerability under various extreme socio-economic conditions and following extreme weather events (Islam & Walkerden, 2014; Malak et al., 2020; Nguyen-Trung et al., 2020). Care and support for family, community and group members are provided through social structures and social networks (Calder et al., 2014). The assistance–financial or material–is often provided by neighbours, friends and relatives who are not affected and their ability. Usually, informal transfers are made by community members, relatives, private philanthropists, religious organisations and other NGOs to the poor and vulnerable groups in the form of cash, goods, food and in-kind support. This type of support helps older people to predict, cope and deal with extreme weather events. They could be ex-ante mitigation and prevention tools and ex-post complementary coping measures or formal social protection programmes for older people.

7.3.1 Volunteer Groups and NGOs

Like government organisations, voluntary groups, organisations and NGOs could also play an effective role in preparing older people and their households to deal with extreme weather events (Hossain, 2020). These services help older people and their households to prepare for extreme weather events by providing information about the magnitude of the events and their impact on the community. Most older participants indicated that they receive news about extreme weather events from the voluntary organisation. This type of support gives older people access to information and builds their anticipatory skills so that they can reduce damage and loss during extreme circumstances. Rashidul said:

> We learn from volunteers. When the weather forecast is broadcast on TV and radio, the volunteers of Red Crescent Bangladesh announce it through loudspeakers along the road. They confirm that there will be cyclones, heavy rains and floods and we head to the shelter.

Some NGOs provided food, water, clothes, and medicines to the elderly and other people in the community. They also prepared shelters with the help of GOs, communicated with vulnerable groups and made them aware of the impact of the extreme events. Shuvasm said:

> ... NGOs help in disasters. They provide rice, flour, a matchbox, flatten rice, water, clothes and sometimes money. For example, I got clothes, a tin to repair my house, and food. An international NGO called CARITAS gave me tin.

Getting the house and home repair materials from NGOs provided significant assistance to the older people and their households in the study area after extreme weather events. Chandresh, a widower from Haritana village, said that after cyclone SIDR in 2007, OXFAM–an NGO–provided tin to construct his house. Shamiul, a 72-year-old fisherman from West Haritana village, said, "... The NGO named Sangram built a house for me after SIDR". Abdullah, an 80-year-old casual labourer from Charlathimara village, said, "I get relief food. It is provided by NGOs." Like other participants, Fakir Shah received aid from NGOs after cyclone SIDR in 2007. He said, "... My house is made of corrugated iron sheets. Sangram, an NGO, provided this corrugated iron sheet and built this small house after SIDR. I live here." Enam Munshi, an 80-year-old widower from Padma village, said, "The NGO provided me with a house. I still live there." Rizeeya, a 70-year-old widow from East Badurtala village, reported, "I have a toilet. It was provided by an NGO after cyclone SIDR [in 2007]".

Although older people have limited access to the formal credit market, microfinance institutions and non-profit organisations offer financial services to them and their households. This type of support enabled them to cope with their financial vulnerability after an extreme weather event. Access to financial services in the form of social protection, savings and loans can help older people get through hard times. "I had once received a loan from an NGO," explained Shuvasm, a small vendor from Haitana Village in the Union. Rashidul from the same village said he received a loan of BDT20,000 from Palli Sanchay Bank. Shamiul, a 72-year-old fisherman from West Haritana village, said, "BRAC [an NGO] gave me a cow as a loan. I am paying back the amount in instalments. Though it belongs to my daughter, I take care of it."

Some participants reported that NGOs initiated employment programmes such as 'food or cash for work' for unemployed older people and their household members after extreme weather events. Shakima Begum, a 60-year-old widow from Padma village, said, "After the cyclone, I worked for [road repair] Sushilon, an NGO, digging the earth." Some of the participants said that various NGOs help them develop saving habits to cope with future economic crises. Hussain, a 67-year-old farmer and casual labourer, said:

> I have some savings. I save BDT50 a week with an NGO called Sangram. When it reaches BDT1000/2000, I can withdraw it. I could use it in case of a disaster or when a relative comes to visit. I have no investment plan other than buying a cow.

Most organisations focus on food relief. However, some organisations generally separate their relief efforts into food and non-food categories. For example, in a 2017 flood study, it was found that government and non-government organisations provided rice, pulses, cooking oil, salt, sugar, and baby cereal, as food items, which were packed into different food packages. The main non-food relief items were corrugated iron sheets, blankets, sarees, lungis, mosquito nets, ropes, family kits, kitchen kits, ORS packages, buckets, mugs, soap, sanitary napkins and washing powder. It is clear that the GOs played comparatively a larger role in providing cash and corrugated iron sheets. In contrast, NGOs were more effective at providing food, water, clothing, and medicine (Hossain, 2020).

7.3.2 Neighbours and Friends

During extreme weather events, when family members and relatives are unable or unwilling to provide financial support to older people, they turn to their neighbours. When neighbours notice that someone is missing breakfast and lunch, they bring them dinner so that they do not go hungry (Erb, 2011). Informal support from friends and neighbours, who are the main carers of older people, is evident during socio-economic crises (Nocon & Pearson, 2000). Participants in the present study indicated that they received help from their neighbours and relatives before, during and after extreme weather events. Some older respondents said that they could find shelter in their neighbours' or relatives' houses during extreme weather events. Rashidul from West Haritana village said:

> My neighbours helped me when I had a problem after the cyclone. For example, they helped me remove trees that had fallen on my house during the last cyclone. I had a cordial relationship with them. When I was sick one day, they all came to help me.

Some felt that mutual help is naturally beneficial in times of extreme weather events when most of them feel hopeless. A 73-year-old farmer and casual worker, Chandra from Charlathimara village received help from their neighbours and relatives during and after the cyclones. He said, "People ask each other for food and where to stay during the disaster." Amena, an 80-year-old widow from Padma village, commented, "Neighbours and relatives help each other after extreme weather events like cyclones. For example, one of my brothers-in-law gave me clothes after the cyclone."

Some reported that they received informal loans that helped them to mitigate vulnerability to extreme weather events at different times. "[...] If possible, they lend me money or goods that I have to pay back later", said Shuvasm, a betel leaf vendor from Haritana village. Kamal, a shopkeeper from Badurtala Bazar, said people who were solvent and less affected distributed cooked food to them for a week after cyclone SIDR in 2007. Some neighbours allowed landless people to live on their unused land in the area. Palkey Begum, for example, lives with her young son on a khas land given to her by her neighbour. She has no land to cultivate or live on. She said:

> ... I do not have any land. [My] husband also had no land or property. My husband's home was Mathbaria [a nearby Upazila in Pirojpur district]. And their ancestral home was Patarghata. After my husband's death, I came here with my child. [...] Here I lead my life by working hard. I live on the khas land [land belonging to the state].

Older people, women and children, were more vulnerable than others in the community. They face many obstacles during the evacuation, on the way to safe places such as shelters, and after extreme weather events. Some said that other young people in the community help them reach a safe place when they find us helpless on the way to the shelter centres. Chandresh, a 77-year-old widower from Hariana Village said:

> ... [though] everyone thinks about themselves at that time [during cyclones], if anyone watches them, they try to help [older people] to take with them [to safe places]. People try to save them [older people] first.

Fakir Shah, 61 years old and a fisherman from Padma village, said he has good relations with his neighbours. There is no quarrel there; he has no chaos or confusion with his neighbours. Whenever possible, neighbours try to help each other and work together in times of extreme events and danger. This is what he did. He said:

> Young people arrange the necessary arrangements and take actions to bring the old people to the shelter. Volunteers first bring the old people to the shelter and keep them in a separate place.

Lack of potable drinking water was a common problem for older people, their families, and the community at large. Almost all stated that they had to use pond water. The government had provided a filtration system to treat the pond water and store it in a tank. They had to fetch the water from the tank and use it for drinking and cooking. Sometimes neighbours helped them to collect water from their tube well. Older people were not physically strong enough to carry the water from the pond because it was very far away. Their neighbours or relatives sometimes helped them to fetch water for them.

Generally, older people rely on their extended families and their communities for their livelihoods. Older people who need outside help to build sustainable livelihoods often turn to moneylenders or pawnbrokers, as well as microfinance institutions and NGOs. Older people face discrimination in accessing savings and loans offered by local microfinance institutions due to their age. They are not excluded from banks because of their age, but rather because of a lack of economic means. Whenever necessary and when they are able to, friends, neighbours and the community provide emergency assistance (Erb, 2011).

7.3.3 Family Members and Relatives

Family members and relatives are seen as a great source of support before, during and after an extreme weather event. Most older people received cash, materials, food and in-kind support from their family members and relatives after the storms. Rashidul said that he had difficulty meeting his living expenses. His sons sometimes help him raise money. He said, "If I say I do not have clothes, they buy them [clothes] for me. They give me some cash. I then use that to buy food and goods." He also mentioned that he cannot wash his clothes. His wife helps him get water to bathe and wash his clothes.

Some explained that although their relatives sometimes help them during and after cyclones, their family members–sons and daughters–support them the most. Seventy-two-year-old Shamiul, a casual fisherman, said, "I am helpless and poor. [...] I am supported by my children [after the extreme events]." Biddut Hossain, a 62-year-old fisherman, said that while fishing in the river, fish jumped at him and injured his eyes. Nearly BDT200,000 was spent on treating his eyes. "My relatives helped me during the cyclone. [...] One of my daughters-in-law came with some rice," he added. Kamal, a 60-year-old shopkeeper from Badurtala village, said he lost everything in his shop during a previous cyclone. His relatives who were in Dhaka sent him clothes, candles, and other essentials. He said, “My maternal uncle-in-law from Dhaka and a relative from Comilla helped me [during the past cyclone].”

It is argued that families, friends, neighbours and the community provide older people with only a shaky economic safety net that is hardly regular, reliable and consistent (Erb, 2011). However, informal social protection through friends and family was seen by most people as a more reliable source of socio-economic security than formal social protection (Heltberg et al., 2007). Most safety nets available to older people were family, friends and mutual solidarity groups, which weakened as the crisis progressed (Browne, 2015). In the context of individual shocks and events, informal and kinship-based social protection such as neighbourhood credit and family support tended to work well (Hebbar & Shebab, 2020).

7.4 Discussion and Conclusion

A well-designed and efficiently managed social protection system for the older population can not only promote age equality but also transform old people into productive members of society (Farid, 2019). Evidence shows that building capacity and resilience for older people and their vulnerable families depends on investing in a more comprehensive social protection system that includes both individual and community programmes with formal and informal measures. There is a need to take stock of a country’s social protection system to strengthen the resilience of the poor and vulnerable to covariate extreme weather events (Bowen et al., 2020). Social protection programmes have been shown to improve a household’s capacity to cope with extreme events by providing support after the shock occurs in three ways: Supporting coping capacity, supporting consumption and reducing food insecurity (Ulrichs & Slater, 2016). The experiences of older people in coastal Bangladesh showed how social protection programmes improve the level of socioeconomic life, health, and food supply of older people during and after extreme weather events which improves their capacity and reduces their vulnerability.

Extreme weather events have many adverse impacts on the vulnerability of older people and their families in coastal Bangladesh because of limited access to support measures and tools to. In the coastal region of Bangladesh, the various social protection programmes help older people to mitigate their vulnerability to extreme weather events. Social protection programmes that respond to shocks can strengthen resilience by enhancing households’ capacity to anticipate, absorb, and adapt to covariate shocks (Hebbar & Shebab, 2020). Protective measures of social protection can help older people reduce their exposure to extreme weather events, improving their ability to withstand the impacts of a climatic disaster. Conversely, social protection coping mechanisms enable older people to adapt to the consequences of extreme weather events (Vakis, 2006). Evidence shows that social protection programmes, both formal and informal, can improve the status and honour of beneficiaries over time. Despite its relatively small allocation, social protection has shown some positive outcomes around the world (Choudhury, 2012; Hasan, 2012; Khan, 2012; Uddin, 2013).

The experiences of older people in coastal Bangladesh suggest that various social protection programmes have played a significant role in the lives of older people and their families in the aftermath of extreme weather events in the region. It is clear that old age and widow allowances make a significant difference to the lives of beneficiaries by providing them with access to food, medical care, and clothing, and by overall strengthening family ties and enhancing the quality of life. These allowances enable respondents to meet their minimum basic needs. Most of the allowance is spent on food, clothing and medicines for older people and their family members. The allowance enables many elderly recipients to increase their food consumption in the household. Older people who receive an allowance can consume more protein-rich food such as fish, milk, meat and other nutritious food than households that do not receive the allowance. It was found that most beneficiaries of the old age allowance spend on food and clothing as well as their daily consumption habits change dramatically after receiving the old age allowance (Uddin, 2013). Further, most beneficiaries have meals three times a day thanks to the old age allowance. It was also found that the food consumed by respondents at each meal had increased since they started receiving the allowance. Two in five older people consumed rice, lentils, fish and vegetables after receiving the allowance.

Access to health care and medications was critical during their illness and when older people were vulnerable to diseases before and after extreme weather. The present study has shown that the recipients of Old Age Allowance and Widow Allowance spend more on necessary medicines and health needs. It is not uncommon for allowance recipients to spend more on health care than those who do not receive allowances. Other studies on the effects of the Old Age Allowance also found that a majority of beneficiaries could afford emergency medications and had some access to visit doctors. Further, older people were able to receive care from family members in case of complications in old age and illness (Hasan, 2012; Rana & Ahmed, 2010; Uddin, 2013).

Old age allowance and widow allowance contribute to improving the dignity of older people in their families and their communities by strengthening the bond with their family members and community members. In most cases, older people receive better food from their family members and more care from their social network due to receiving allowances. Since they can support themselves financially, in some cases they can be financially independent and help other family members when they are in need. As a result of social protection programmes, they gain more self-esteem and a greater sense of honour and importance in the eyes of family members, friends, local politicians, and society in general. In addition, the negative attitude of the community, peers and local elites towards them that they are a burden to society decreases (Hasan, 2012). In another study, it was found that social protection such as Old Age Allowance in Bangladesh significantly improves the social relations of elderly beneficiaries (Karim & Hossain, 2013). After receiving the old age allowance, recipients' relationships with their sons, daughters, relatives and neighbours improved significantly compared to the time before receiving the allowance. It seems that social protection prevents the deterioration of social relations of those who previously had good social relations, and it helps to improve the social relations of those who did not have such good social relations.

Again, access to cash through getting an allowance has been found to significantly boost their self-esteem, especially in times of crisis. It enables them to participate in family decision-making. In addition, the old age allowance creates trust among neighbours and shopkeepers. As a result, people in the community trust them to provide them with the goods they need in an emergency, in the hope that they will be able to recoup the costs once they receive the allowance. In a study conducted, nearly 97% of respondents reported that their dignity within the family had increased as a result of this programme (Uddin, 2013). In addition, 60% of respondents reported that their family members now care more about them than before, and 30% are now able to actively participate in decision-making processes within their family. In addition, social protection programmes protect beneficiaries and their families from shocks and extreme situations. Many reported that they had a positive experience with store owners in an emergency and the acceptance of welfare beneficiaries by officials and local people had increased.

The social protection programmes have also improved the living conditions of older people in the area. The older people who receive old age allowance or other types of financial support live better than those without these benefits. The experiences of beneficiaries suggest that their standard of living improved as a result of receiving the allowance. They can now meet their basic needs and can live with family under one roof. The allowance programme has been found to be one of the most effective ways to help older people to save some money that they can invest in various income-generating activities such as raising cattle, chickens or ducks or in small businesses (Hasan, 2012).

The NGOs, neighbours, relatives and family members provide informal social protection. In order to receive help in an emergency, whether material or financial, a reliable social network is crucial. It was found that most help is provided by family members and relatives, followed by friends and neighbours. These types of social support help older people to recover immediately from the effects of extreme weather events. Although the range of sources of support available to them may vary, they are helpful after extreme weather events. A study conducted to compare the positions of governmental and NGOs found that nearly half of the respondents received food items, about one-fourth received clothing, 37% received cash and 8% received other types of assistance from government organisations; and, from the NGOs, about three-fourth of them received food, 52% received pure drinking water, 40% received clothes, and one-fourth had access to other facilities (Hossain, 2020). Also, most older people receive some form of support from their families, including those who have migrated to other parts of Bangladesh or abroad (Khondker et al., 2013). Mutual support through social capital helps older people cope with adverse situations after a cyclone disaster. Social networks include not only family members (bonding social capital) but also friends, neighbours and community groups including religious and charitable associations (bridging social capital), as well as mutual aid and charity agreements (Islam & Walkerden, 2014; Malak et al., 2020; Nguyen-Trung et al., 2020).

Mutual social support helps older people cope with crises after a cyclone. For example, they provided food to neighbours immediately after a cyclone when institutional help was not available and helped each other rebuild their homes. Similar findings are found in other studies.). Other studies in coastal Bangladesh and Vietnam have shown that mutual support through bonding and bridging social networks played a crucial role throughout the short- and long-term recovery process after extreme weather events (Islam & Walkerden, 2014; Nguyen-Trung et al., 2020; Paul, 2012, 2015).

Nevertheless, the amount of old age allowance plays an important role in the wellbeing of the beneficiaries. Older people who think the current amount is sufficient will have a strong sense of security, but if they think it is insufficient or insignificant, the sense of security will be weaker. In this study, it was also found that the amount of money given as an allowance was not enough to cover the total family expenses for older people who have extended families. Although the amount they received was small, it was still useful for them and they would likely face many difficulties when they no longer received the allowance. In a previous study, it was also found that the amount was insufficient for the recipients. Half of the respondents said that the current amount of allowance was insignificant. Thirty-eight per cent said it was insignificant. Only 12% thought it was moderate. None of the respondents considered it very significant or significant (Hasan, 2012). Older people in coastal Bangladesh face a variety of risks and vulnerabilities. However, their impacts are sometimes more severe than those of the general community. The negative impacts of extreme weather events are more detrimental to them than to those who are better off financially, physically, psychologically, and in terms of long-term human development capacity. A social protection programme can therefore play an important role in helping older people and their families to address their vulnerability and cope with losses following extreme weather events (Hasan, 2012).

The vulnerability of older people is argued to result from a combination of the cumulative effects of extreme events and processes that have taken place during their lifetime. There is a constant and contradictory interaction between resources, risks, threats and socio-economic conditions, extreme weather events, social networks and the support system available in the community. The vulnerability of older people is a dynamic process rooted in both internal and external livelihood and geological factors. Besides their efforts, a range of social protection programmes, both formal and informal, can therefore help to address their vulnerability by strengthening their capacities. The three main contributions of social protection identified in the experiences of older people in coastal Bangladesh are: it improves the socio-economic conditions of older people and their households, mitigates their vulnerability and increases their capacity. Providing housing opportunities, improving relationships with family, friends and relatives, and facilitating income-generating activities improve the socioeconomic status, consumption and food security of older people and enable them to adapt to extreme weather events in the coastal region of Bangladesh.

References

Ahmed, A. U. (2005). Adaptation options for managing water related extreme events under climate change regime: Bangladesh perspectives. *Climate Change and water Resources in South Asia,* Balkema Press, Leiden, 255–278.

Barrientos, A. (2011). *Non-contributory pensions and social protection. In non-contributory pensions and social protection* (Working Paper No. BWPI WP 15).

Bastagli, F., & Holmes, R. (2014). *Delivering social protection in the aftermath of a shock: Lessons from Bangladesh, Kenya, Pakistan and Viet Nam.* (ODI report).

Begum, S. (2012). Overview of the Old Age Allowance Programme in Bangladesh. In Social Protection for Older Persons: Social Pensions in Asia (pp. 187–213). Asian Development Bank. https://doi.org/10.13140/2.1.2378.5769

Béné, C., Newsham, A., Davies, M., Matthes, F., & Mitchell, T. (2014). *Social protection and climate change* (OECD Development Co-operation Working Papers No. 16). OECD Publishing.

Bowen, T., Devereux, S., Sabates-Wheeler, R., & Slater, R. (2020). *Adaptive social protection: Building resilience to shocks.* World Bank.

Browne, E. (2015). *Social protection: Topic guide.* GSDRC, University of Birmingham.

Calder, R., Tanhchareun, T., & Crescent, J. M. (2014). *Informal social protection: Social relations and cash transfers.* Australian Government-Department of Foreign Affairs and Trade.

Canelas, C., & Niño-Zarazúa, M. (2022). *Social protection and the informal economy: What do we know?* United Nations Economic and Social Commission for Asia and the Pacific (ESCAP). Retrieved December 2, 2025, from https://www.unescap.org/kp/2022/social-protection-and-informal-economy-what-do-we-know

Chandra, D., Khan, A. A., & Golder, B. (2017). Widow allowance scheme and women empowerment in Bangladesh: Role and implications. *Journal of Science and Technology, 51*, 60.

Choudhury, L. A. (2012). *Social safety nets in Bangladesh, Volume 2: Ground realities and policy challenges.* Power and Participation Research Centre; United Nations Development Programme.

Cull, R., Demirgüç-Kunt, A., & Morduch, J. (2014). *Financial inclusion and development: Recent impact evidence* (CGAP Focus Note).

Davies, M., Oswald, K., Mitchell, T., & Tanner, T. (2008). *Climate change adaptation, disaster risk reduction and social protection: Briefing note.* Institute of Development Studies.

Erb, S. (2011). *A study of older people's livelihoods in Bangladesh.* HelpAge International and Cordaid.

Farid, S. (2019). Reimagining social protection for older people. *The Daily Star.* Retrieved July 28, 2021, from https://www.thedailystar.net/opinion/society/news/reimagining-social-protection-older-people-1807489

Fiorillo, C., & Matin, I. (2009). *Study of the first phase of the 100-day employment generation programme.* National Food Policy Capacity Strengthening Programme and BRAC.

Grosh, M., Andrews, C., Quintana, R., & Rodriguez-Alas, C. (2011). *Assessing safety net readiness in response to food price volatility* (SP Discussion Paper No. 1118). World Bank.

Haider, M. Z., & Mahamud, A. (2017). Beneficiary selection and allowance utilization of social safety net programme in Bangladesh. *Journal of Human Rights and Social Work, 2*(1), 45–51.

Hasan, M. M. (2012). *Effectiveness of old age allowance program in Bangladesh: A study on the demand side dynamics* (Master in Public Policy and Governance Program). Department of General and Continuing Education, North South University, Bangladesh.

Hebbar, M., & Shebab, S. (2020). *Shock-responsive social protection in Bangladesh.* OPM and UK aid.

Heltberg, R. (2007). Helping South Asia cope better with natural disasters: The role of social protection. *Development Policy Review, 25*(6), 681–698.

Hossain, B. (2020). Role of organizations in preparedness and emergency response to flood disaster in Bangladesh. *Geoenvironmental Disasters, 7*(1), 33.

Islam, K. M. (2007). *Poverty reduction in Bangladesh: A qualitative analysis of the "Allowance Scheme for Widowed and Husband-deserted Distressed Women"* (Master's thesis). Bodo University College. Bodø, Norway.

Islam, R., & Walkerden, G. (2014). How bonding and bridging networks contribute to disaster resilience and recovery on the Bangladeshi coast. *International Journal of Disaster Risk Reduction, 10*, 281–291.

Kabir, R., et al. (2016). Climate change and public health challenges in Bangladesh: A review. *Journal of Health, Population and Nutrition, 35*(1), 1–12.

Karim, M. F., & Mimura, N. (2008). Impacts of climate change and sea-level rise on cyclonic storm surge floods in Bangladesh. *Global Environmental Change, 18*(3), 490–500.

Karim, M. R., & Hossain, M. F. (2013). Impact of old age allowance on social relations of the elderly in rural Bangladesh. *Jagannath University Journal of Social Sciences, 1*(1–2), 1–2.

Kasseeah, H., & Tandrayen-Ragoobur, V. (2011). Assessing the formal social protection system in Mauritius. *The International Journal's Research Journal of Social Sciences and Management, 1*(6), 82–94.

Khan, N. J. (2012). *An assessment of the widow allowance programme in Bangladesh: The supply side perspectives*. North South University.

Khondker, B. H., Knox-Vydmanov, C., & Vilela, A. (2013). *Old age social protection options for Bangladesh*. Retrieved April 27, 2021, from https://www.researchgate.net/publication/277588545_Old_Age_Social_Protection_Options_for_Bangladesh

Lloyd-Sherlock, P. (2006). Social protection for older people: A review of the literature. *Ageing & Society, 26*(4), 517–536.

Malak, M. A., Sajib, A. M., Quader, M. A., & Anjum, H. (2020). "We are feeling older than our age": Vulnerability and adaptive strategies of aging people to cyclones in coastal Bangladesh. *International Journal of Disaster Risk Reduction, 48*, 101595.

Nguyen-Trung, K., Forbes-Mewett, H., & Arunachalam, D. (2020). Social support from bonding and bridging relationships in disaster recovery: Findings from a slow-onset disaster. *International Journal of Disaster Risk Reduction, 46*, 101501.

Nocon, A., & Pearson, M. (2000). The roles of friends and neighbours in providing support for older people. *Ageing and Society, 20*(3), 341–361.

Norton, A., Conway, T., & Foster, M. (2001). *Social protection concepts and approaches: Implications for policy and practice in international development* (Working Paper 143). ODI.

Oduro, A. D. (2010). *Formal and informal social protection in sub-Saharan Africa* (Master's thesis). University of Ghana. Legon, Accra, Ghana.

Paul, B. K. (2012). Factors affecting evacuation behaviour: The case of 2007 Cyclone Sidr Bangladesh. *The Professional Geographer, 64*(3), 401–414.

Paul, B. K. (2015). *Disaster vulnerability and resilience in Bangladesh*. Springer.

Rana, A. K. M. M., & Ahmed, S. M. (2010). *Impact of Old Age Allowance on Health-Related Quality of Life Among Elderly Persons in Bangladesh*. Retrieved July 14, from https://www.eldis.org/document/A42274

Tenzing, J. D. (2020). Integrating social protection and climate change adaptation: A review. *Wiley Interdisciplinary Reviews: Climate Change, 11*(2), e626.

Uddin, M. A. (2013). *Social safety nets in Bangladesh: An analysis of the impact of old age allowance programme*. Brac Institute of Governance and Development (BIGD), Brac University.

Ulrichs, M., & Slater, R. (2016). *How can social protection build resilience? Insights from Ethiopia, Kenya and Uganda* (BRACED Working Paper). ODI.

Vakis, R. N. (2006). *Complementing natural disasters management: The role of social protection* (Social Protection Discussion Paper 0543). World Bank.

Chapter 8
Conclusion and Policy Recommendations

8.1 Introduction

Extreme weather events adversely affect people by making them vulnerable to socio-economic conditions. Older people are the most vulnerable to the impact of extreme weather events and have a low adaptive capacity to cope with severe climate disasters and their impact on their socio-economic conditions. This book has examined the impact of extreme weather events on the vulnerability of older people and explored how social protection programmes respond to their vulnerability. It has attempted to address the following questions: (a) Why are older people in coastal Bangladesh vulnerable to extreme weather events? (b) How do extreme weather events exacerbate the vulnerability of older people in coastal Bangladesh? and (c) How do social protection programmes address the vulnerability of older people in coastal Bangladesh? These questions were examined using the lived experiences of many older people in the coastal region of Bangladesh, which is subject to recurrent extreme weather events.

The concept of vulnerability is understood differently across disciplines, but simply put, it means being exposed to threats and having the limited adaptive capacity to deal with these threats effectively. Chambers (1989, p. 1) defines vulnerability as "exposure to contingencies and stress, and difficulty in coping with them". The contingencies of older people here include unexpected extreme weather events causing the 'irreversible loss or sale of assets', loss of employment and deterioration of health. This means that the vulnerability of older people is the lack of capacity to cope with the consequences of extreme weather events. As a result, they would suffer the loss of property or assets, which could put them in vulnerable conditions. The definition has an external aspect of hazards and shocks to which individuals or households are exposed, and an internal component, namely the lack of resources to cope with the risks, threats or hazards.

According to Moser (1998), the more resources people have, the less vulnerable they are. These resources are diverse and include human capital, psychological

M. I. Ali et al., *Extreme Weather Events and Vulnerability of Older People*,
https://doi.org/10.1007/978-3-032-11256-9_8

capital, social capital, information capital, economic capital, political capital and natural capital. When older people lack the reserves and resources or the resilience and capacity to minimise challenges and risks or threats, they become vulnerable to extreme (weather) events. As vulnerability is multidimensional, coming from different sources and varying according to context, it is a challenge to find a common or consensual definition of the vulnerability of older people. Vulnerability is not only due to external sources (e.g. extreme weather events) but also to insecure conditions that are deeply rooted in social, political and economic systems. In this book, the vulnerability of older people to extreme weather events is defined as "a condition resulting from the consequences of the cumulative impact of extreme weather events combined with a lack of capacity to access socio-economic resources and social protection programmes."

The vulnerability of older people to extreme weather events consists of the negative effects of exposure, sensitivity and ability. Grundy (2006) argues that vulnerable older people are those whose reserve capacity is below the appropriate level to cope with the challenges (such as the consequences of extreme weather events) they face. This 'reserve' includes income and material resources, family and social support, and health status. Therefore, the vulnerability of older people in a socio-economic context is a complex, dynamic, challenging, multidimensional and pre-existing condition that depends on both individual characteristics–age, gender, education, health–and social and economic factors such as employment, income, access to credit, social connections and support (Daddoust et al., 2018). The vulnerability of older people thus refers to the functionality of older people and their conditions due to their ability to respond to risks or shocks in an unstable situation during extreme weather events. These functions include primary wellbeing, livelihood and resilience, self-protection, social protection and socio-political and institutional networks (Birkmann, 2006). Social protection can be an important strategy to cushion the impact of shocks and stress through income replacement and strengthening resilience by reducing exposure to extreme weather events, providing incentives/assistance to retrofit homes to withstand tropical cyclones, or implementing public works programmes to restore income to cope with the impact of extreme events. Social protection is commonly understood as all public and private initiatives that provide income or consumption transfers to the poor, protect the vulnerable from livelihood risks and improve the social status and rights of the marginalised, with the overall aim of reducing the economic and social vulnerability of poor, vulnerable and marginalised groups (Devereux & Sabates-Wheeler, 2004).

The experiences of older people in coastal areas of Bangladesh showed that they have been highly vulnerable to extreme weather events in the past. They believe that the same or more brutal conditions will recur due to their poor socio-economic condition and lack of adequate support system. Every extreme event, especially cyclones and floods due to storm surges, destroyed or significantly damaged the livelihoods of older people and negatively affected their lives, making them vulnerable. However, social protection programmes (formal and informal) helped them cope better with the impacts of extreme weather events.

8.2 Sources of Vulnerability

The vulnerability of older people is multidimensional and shaped by a variety of factors, from extreme weather events to the socioeconomic conditions of older people. Climate events, directly and indirectly, affect the income, employment, expenditure and credit liabilities in housing, assets, food and nutrition, water and sanitation, and health of older people and their households in coastal Bangladesh. The origin of older people's vulnerability dates back to about 150 years ago when their grandparents or ancestors started clearing the jungle in the wetlands and settled here to forage, live and grow crops. Their vulnerability was passed down from generation to generation with no permanent solution. Being close to the river and the sea, these settlements have become more vulnerable to extreme weather events. The vulnerability of their settlements to extreme weather events, especially cyclones and floods, is related to three mutually reinforcing factors: location, the likelihood of an extreme climate event occurring, and the extent to which their structures can be damaged. Their houses are one of the main causes of vulnerability because they are built of cheap and poor building materials such as mud, corrugated iron and tin. Many of these houses are light, typically wooden frame structures, which makes them particularly vulnerable to cyclones.

Older people living in absolute poverty lacked the means to cope with extreme weather events, increasing their vulnerability. In addition, the livelihoods of older people from low-income families in coastal areas were associated with endemic poverty, and their vulnerability to extreme weather events may have increased due to lower savings or economic difficulties. Again, older people in the region were found to have fewer employment opportunities. Some jobs were not accessible to older people due to their poor physical strength and limited mobility. The grandparents and parents of older people owned some properties, some of which they lost due to extreme weather events or other social or family crises. Some older people who inherited property from their grandparents or parents were less vulnerable for a few years but had to sell it to survive various extreme weather events, which pushed them into asset vulnerability. This background hindered their upbringing, education and greater job opportunities, thus making them vulnerable in old age.

In general, most older people in the coastal region were socio-economically dependent to some extent on their family members and the social support system, as their economic incapacity and lack of resources made them vulnerable to extreme events. Before and after extreme weather events, when the social support system was inadequate, they depended on the children for daily needs such as food, clothing and medicines.

8.3 Factors that Exacerbate Vulnerability

Relatively high unemployment and underemployment significantly reduce the income of older people and their families. Post-cyclone environmental hazards restricted their access to work, which negatively affected their income and expenditure. The availability of work was found to decrease sharply after the extreme events due to the destruction of sources of employment such as agriculture, fishing and informal work in the forests. Older people and their household members had little or no work after the cyclones and floods. Again, access to work was limited by various environmental and structural hazards such as damage to bridges, roads, houses, factories and communication networks, and loss of equipment for self-employed workers. In addition, as they were physically weak, they were sometimes underpaid in the aftermath conditions. These factors contributed to the financial vulnerability of older people in coastal Bangladesh.

Despite their limited access to credit markets, some older people borrowed money from various sources such as banks, NGOs and moneylenders to rebuild their destroyed houses or build a new one, cover medical expenses, buy food, clothes, agricultural and fishing equipment, grow crops and catch fish. The loss of employment and income, the increase in commodity prices and the loss of purchasing power and credit liabilities left the older people of the region in vulnerable circumstances.

In addition to the loss of employment and income, extreme weather events destroyed various assets, properties and the land of older people and their families in the coastal region almost every year. Older people and their households lost land, houses, fishing boats, nets and machinery, production facilities, stocks of small businesses and resources for self-reliance. The severe cyclones and floods damaged their fishponds and fishing grounds, and they lost their family's livestock and birds. Further, the cyclone shelters were not sufficient to serve all the people living in the area and most of the shelters were far from the homes of older people. It was also found that the structural conditions, atmosphere and facilities of the cyclone shelters were not suitable or adequate for older people.

Among other vulnerabilities, elderly people in the area were also found to suffer from food insecurity. Following the extreme weather events, particularly cyclones and floods, food and nutrient shortages worsened, exacerbating food insecurity among older people. There were several causes of food insecurity after the extreme events, which include the unavailability of food because supply systems were affected by the destruction of communication infrastructures, the loss of local production and access to food due to lack of income and savings, and price increases. The food intake of older people in the study area was affected by extreme weather events and their ability to buy food or the availability and accessibility of food within the family. There were days when they had nothing to eat. Many older people had no access to nutritious food such as meat, fish, eggs and fruits except when provided by a close relative, a neighbour or at a religious festival during the year.

Most older people in the study area were vulnerable to diseases and ailments after extreme weather events. They were unable to obtain healthy and nutritious food and adequate medicines due to their financial inability. They were also dependent on the local pharmacy to obtain medicines in case of illness, as health services in the area were inadequate. In addition to physical health problems, there were also mental health problems among older people in the coastal region. When they recalled the suffering and grief they had experienced during previous extreme weather events, the older people seemed to relive the trauma. They reported that the loss of family members, jobs, houses, assets, farmland and livestock in the past, led to mental health problems among them.

8.4 Role of Social Protection in Addressing Vulnerability of Older People

Extreme weather events have many adverse impacts on the vulnerability of older people and their families in coastal Bangladesh, as they have limited access to support and tools to address their vulnerability. The social protection programmes (SP) such as the Old Age Allowance and Allowances for the Widowed, Deserted and Destitute Women help older people to minimise their vulnerability to extreme weather events. Social protection programmes that respond to shocks can strengthen resilience by improving households' ability to anticipate, absorb and adapt to covariate shocks. Social Protection programmes make a significant difference in the lives of beneficiaries by providing access to food, medical care and clothing, and overall strengthening family ties and improving quality of life. After receiving the allowances, most older people were able to meet their minimum basic needs. Most of the allowance was spent on food, clothing and medicines for themselves and their family members. The allowance enabled many elderly recipients to increase their food consumption in the household. Older people who received the allowance were able to consume more protein-rich foods such as fish, milk, meat and other nutritious foods compared to those who did not receive the allowance. The programmes helped to improve the dignity of older people in their families and their communities by strengthening their bonds with their family members and community members. In most cases, older people received better food from their family members and more care from their social network because of the allowances. Since they were able to support themselves financially, in some cases they were financially independent and helped other family members when they were in need. Again, access to cash through social protection has been shown to increase participation in family decision-making. In addition, the Old Age Allowance created faith among neighbours and shopkeepers. As a result, people in the community trusted them to provide the goods they needed in an emergency in the hope that they would be able to recoup the costs once they received the allowance. Many beneficiaries indicated that their standard of living had improved as a result of receiving the allowance. They could now meet their basic needs and live with their family under one roof.

Social protection–formal and informal–can play an important role in reducing older peoples' vulnerability to the effects of extreme weather events and socio-economic crises. Beneficiaries of formal social protection programmes were found to receive long-term benefits, while informal social protection programmes responded to shocks and provided short-term benefits.

In sum, this book explored the intersection of three aspects of the vulnerability of older people - sources of vulnerability, factors that exacerbate vulnerability to extreme weather events, and the role of social protection in addressing older people's vulnerability in coastal regions in coping with the impacts of extreme climate events. The experiences of older people have shown that there are strong links between resources, impacts and social protection that make older people in the region vulnerable to extreme weather events. By considering the three interrelated issues together, this book provides a unique and comprehensive understanding of why and how older people are vulnerable to extreme weather events the vulnerability of older people to frequent extreme weather events.

Bangladesh has initiated various short- and long-term actions and plans to mitigate the vulnerability of older people and excluded groups under the " leave no one behind" theme of the Sustainable Development Goals (SDGs). The findings presented in this book could help policymakers and researchers better understand the impacts of extreme weather events in the future and the policy responses needed to address the consequences of similar climate disasters. Specifically, the findings could help develop and refine policies to address the vulnerability of older people and marginalised groups to extreme weather events in coastal regions. Importantly, this could be instrumental in refining and expanding the social protection programmes to address the vulnerability of older people and other vulnerable groups in coping with extreme weather events.

8.5 Limitations of the Book

This book explores the vulnerability of older people to extreme weather events and the contribution of social protection to reducing this vulnerability. However, in this study, only older people were interviewed to obtain data and information about their experiences and needs before, during and after the extreme weather events. The household members of older people and other community members in different age groups were not included in the study. Therefore, the results presented in this book do not include detailed information about other household members' needs and experiences of the extreme weather events in the region.

Further, this book covers the experiences of older people in a particular Union of the coastal region of Bangladesh. It does not cover all coastal communities and other parts of the country that are affected by extreme weather events and may increase the vulnerability of older people.

In-depth interviews and field notes were used in this study only to collect qualitative data and information from a specific age group of older people. This study did not use a mixed methodological approach, and the interviews were manually transcribed in Bengali and then translated into English, which may limit the comprehensive assessment of older people's vulnerability in the coastal community.

8.6 Policy Recommendations

This book explored the sources and socio-economic conditions of the vulnerability of older people to extreme weather events in coastal Bangladesh. It also highlighted the contribution that social protection programmes can make to effectively address these vulnerabilities. The experiences of older people showed that they had limited access to family and local resources and social networks, which hindered the effective mitigation of their vulnerabilities. Addressing these issues requires that the identified problems and constraints of older people need to be incorporated into the individual and comprehensive policy reform frameworks and future planning–coastal planning, disaster programmes and infrastructure development. Adaptation measures that can be taken to reduce older people's vulnerability to extreme weather events include:

- Non-governmental organisations' staff can reduce the negative impact of extreme weather events by informing older people about their vulnerability in the region.
- Vulnerability directories based on the socio-economic and demographic data of older people, as well as vulnerability mapping, can help policymakers, emergency supervisors and technical assistance organisations to identify vulnerable older people during extreme weather events.
- All stakeholders, including the public and private sectors, communities, civil society and non-governmental organisations, must take responsibility and work together to strengthen the resilience of people and infrastructure.
- Governments can allocate resources to long-term and systemic resilience strategies and strategic actions for older people, balancing national and local priorities.
- National and international authorities should allocate necessary resources to resilience-building initiatives in areas most vulnerable for older people to extreme weather events.

The competent authority of Bangladesh may consider the following recommendations to strengthen the capacity of older people and their households to mitigate their vulnerability to extreme weather events.

8.6.1 Infrastructure Development and Maintenance

The construction and management of buildings, roads, power and telecommunication lines, drainage and sewerage, and waste disposal are very difficult and vulnerable to climate change-related disasters. Cyclones in combination with floods have serious impacts on infrastructures and thus on livelihoods (Rahman et al., 2015).

Thus, timely development and proper maintenance of infrastructure are one of the first necessities in the community to mitigate the vulnerability of older people to extreme weather events. As the coastal region is highly vulnerable to extreme weather events, the policy authority may consider the following infrastructural developments immediately and maintain them properly before the extreme weather seasons.

8.6.1.1 Construction and repair of the dam

The dam is the first line of protection for the elderly and their households from extreme cyclones and floods. The construction of earthen dams for flood control, irrigation and drainage has always been known in Bangladesh, but these cannot solve the flood problem effectively and permanently as they are easily vulnerable to rain splash and the flow of flood water. It is, therefore, necessary to repair the dam immediately when it is damaged or broken due to extreme cyclones and floods and to monitor its proper maintenance.

8.6.1.2 Roads and pathways

The vulnerability of older people to extreme weather events may be exacerbated by the lack of proper road networks and transport infrastructure. During the rainy season, they are destroyed, and the mud makes it difficult to move around. Roads and pathways are essential for the evacuation of older people during extreme weather events. The relevant local government and Union Council may take appropriate measures to construct and repair roads and pathways for the remote older villagers to move safely into the shelters and a safe place during cyclones and floods.

8.6.1.3 Cyclone and flood shelters

Cyclone shelters are considered comparatively safe for older people during extreme weather events. However, their capacity is less than what is needed in the community and the shelters are not safe for them. It is evident that the shelter conditions and distances, the lack of infrastructure facilities and their deterioration pose a problem for nearby households when deciding whether or not to seek cyclone shelters. With the help of donor agencies and local elites, the relevant authorities should construct more cyclone shelters so that older victims of extreme weather events can seek shelter in a safe place during extreme weather events.

8.6.1.4 Housing

To reduce the vulnerability of older people and their households, it is essential to support the construction of more robust houses in coastal villages. Most houses on the coast are not strong against cyclones and storm surges. In addition, many elderly people cannot afford to build sturdy houses to protect themselves from strong cyclones and storm surges, and many of them do not seek shelter in cyclone centres. This is due to inadequate shelter centres, the lack of age- and gender-appropriate environments in the centres, and inappropriate communication systems (both roads and vehicles) between residents and shelter centres. In addition, some locals believe that they should not leave their homes during a disaster for fear of misfortune. Therefore, stable shelters are needed for the elderly in coastal households, which would help minimise damage and losses. The government, in collaboration with domestic and foreign NGOs and donors, can provide better shelter to the locals through long-term, interest-free loans.

8.6.2 Creating Employment Opportunities for Older People

Creating employment opportunities for older people and their households is important to generate income and improve their ability to recover from the loss and damage following extreme weather events. Most older people living in coastal communities have limited access to employment opportunities due to a lack of age-appropriate jobs. The government can therefore create employment-intensive facilities such as fish processing factories, expand tourism and promote other income-generating activities such as tailoring, sewing for older women, poultry farms and small handicrafts in coastal areas.

8.6.3 Give Preference to Older People in the Distribution of Relief

Evidence suggests that older people are often overlooked by emergency response systems during extreme weather events, exacerbating their vulnerability and limiting their ability to respond and adapt. All phases of emergency management and response need to consider the specific needs and challenges faced by older people in extreme weather events. Older people are more vulnerable than any other age group. Older people have little physical and mental strength to move to the central relief distribution points. Therefore, the relevant ministries and local governments could take appropriate measures to distribute relief goods in such a way that the elderly can receive sufficient help and support. A special group could be formed to distribute relief goods for the elderly in the cyclone shelters and from door to door.

8.6.4 A Community Centre for the Elderly

A community centre for older people could facilitate gatherings where they can share their views and experiences on the impact and vulnerability of extreme weather events and interact with local authorities. The local government could play the role of initiating this programme with the help of NGOs and local youth.

8.6.5 Include All Older People in Formal Social Protection Programmes

In Bangladesh, formal social protection for older people included both contributory and non-contributory schemes. The experiences in the coastal region showed that not all older people who were eligible for formal social protection received benefits. As a result, they were more vulnerable than those who received social protection. To address this, the government can include all eligible older people who are poor and vulnerable in the formal social protection programmes, which will help them somehow cope with their financial crisis after extreme weather events.

References

Birkmann, J. (Ed.). (2006). *Measuring vulnerability to natural hazards: Towards disaster resilient societies*. United Nations University Press.

Chambers, R. (1989). Editorial introduction: Vulnerability, coping and policy. *IDS Bulletin, 20*(2), 1–7.

Daddoust, L., Khankeh, H., Ebadi, A., Sahaf, R., Nakhaei, M., & Asgary, A. (2018). The social vulnerability of older people to natural disasters: An integrative review. *Health in Emergencies and Disasters, 4*(1), 5–14.

Devereux, S., & Sabates-Wheeler, R. (2004, October). *Transformative social protection* (Institute of Development Studies (IDS) Working Paper, October). https://www.ids.ac.uk/

Grundy, E. (2006). Ageing and vulnerable older people: European perspectives. *Ageing and Society, 26*(1), 105–134.

Moser, C. (1998). The asset vulnerability framework: Reassessing urban poverty reduction strategies. *World Development, 26*(1), 1–19.

Rahman, S., & Rahman, M. A. (2015). Climate extremes and challenges to infrastructure development in coastal cities in Bangladesh. Weather and Climate Extremes, 7, 96–108.

Appendices

Appendix 1: Old Age Allowance

Criteria for selecting candidates:

a. Citizenship: The candidate must be a permanent citizen of Bangladesh.
b. Age: Preference should be given to the oldest person.
c. Health condition: The person who is physically disabled, i.e., completely disabled, should be given the highest priority.
d. Socio-economic status:

    i. In the case of financial conditions: The destitute, the displaced and the landless should be given priority in order.
    ii. In the case of social conditions, Widows, divorcees, widowers, childless, separated from their families should be given priority in order.

e. Ownership of land: Priority should be given to the landless person. In this case, if the amount of land of a person has other than a house is 0.5 acres or less, then he will be considered landless.

Eligibility and conditions of receipt of allowances

a. must be a permanent resident of the concerned area;
b. Must have birth registration / national identity number;
c. The age should be a minimum of 65 years for men and a minimum of 62 years for women.[1]
d. The annual average income of the candidate should be not less than BDT 10,000 (ten thousand);

[1] In determining the age, national identity card, birth registration certificate, SSC / equivalent examination certificate should be considered. In this case, if there is any dispute, the decision of the concerned committee will be considered final. The age fixed by the government from time to time must be considered.

M. I. Ali et al., *Extreme Weather Events and Vulnerability of Older People*,
https://doi.org/10.1007/978-3-032-11256-9

Table A.1 Statistics on old age allowance

Coverage			The budget			Details of service		
Earlier	Start time	Currently	Earlier	Start time	Currently	Earlier	Start time	Currently
4.03 lakh people	1997–98	57.01 lakh people (2022–23)	BDT 12.50 crore	1997–98	BDT 3444.54 crore	3 months experimental at the rate of BDT 100 per month	1997–98	An allowance of BDT 500 per month is given to the person (2022–23)

Sources Department of Social Services, Ministry of Education, Bangladesh

e. Must be elected by the Selection Committee.

Ineligibility of receiving allowances

a. If a government employee is a pensioner;
b. If the VGD card holder is a poor woman;
c. If regular government grants/allowances are received in any other way;
d. If financial grants / allowances are received regularly from any non-governmental organization / social welfare organization (Table A.1).

Appendix 2: Allowance for Widows and Destitute Women

Standards of Candidate Selection:

a. Citizenship: The candidate must be a permanent citizen of Bangladesh.
b. Age: Age should be above 18 (eighteen) years. But, priority should be given to the most senior women.
c. Health status: The highest priority should be given to the person who is physically disabled or completely non-functioning.
d. Socio-economic status:
    i. In case of financial status: destitute, the refugees and the landless should be given priority in a gradual manner.
    ii. In the case of social conditions: Those who are childless, unhealthy and isolated from families should be given more priority.
    iii. Land Ownership: A Landless candidate has to give priority. He will be considered landless if he has 0.5 acres or less of land except for household.

Terms and conditions for the beneficiary:

i. Be a permanent resident of the area concerned;

ii. Must have birth registration / national ID card;
iii. Priority shall be given to the elderly, helpless and afflicted widow or the women who are deserted by their husband;
iv. A woman who is poor, helpless, almost landless, widowed or homeless and has two children below the age of 5, will be prioritised in getting the allowance;
v. Among the impoverished, the needy, the widow, and the abandoned wife, the disabled and the sick will get more importance in getting allowance.
vi. Annual average income of the candidate must be BDT less than 12,000/- taka.
vii. Must be selected by the selection committee (Table A.2).

Table A.2 An overview of the allowance for the Widow and Husband Deserted Women programmes

Coverage			Budget			Service details		
Beneficiary at the beginning	Beginning time	At present	At beginning	Beginning time	At present	At beginning	Beginning time	At present
4.3 lakh people	1998–99	24.75 lakh people (2022–23)	40.31 core BDT	1998–99	1495.40 core BDT	100 BDT was the monthly allowance	1998–99	500 BDT is a monthly allowance (2022–23)

MIX
Papier aus verantwortungsvollen Quellen
Paper from responsible sources
FSC® C105338

FSC
www.fsc.org

If you have any concerns about our products,
you can contact us on
ProductSafety@springernature.com

In case Publisher is established outside the EU,
the EU authorized representative is:
Springer Nature Customer Service Center GmbH
Europaplatz 3, 69115 Heidelberg, Germany

Printed by Libri Plureos GmbH
in Hamburg, Germany